SpringerBriefs in Electrical and Computer Engineering

Control, Automation and Robotics

Series Editors

Tamer Başar, Coordinated Science Laboratory, University of Illinois at Urbana-Champaign, Urbana, IL, USA

Miroslav Krstic, La Jolla, CA, USA

SpringerBriefs in Control, Automation and Robotics presents concise summaries of theoretical research and practical applications. Featuring compact, authored volumes of 50 to 125 pages, the series covers a range of research, report and instructional content. Typical topics might include:

- a timely report of state-of-the art analytical techniques;
- a bridge between new research results published in journal articles and a contextual literature review;
- a novel development in control theory or state-of-the-art development in robotics;
- an in-depth case study or application example;
- a presentation of core concepts that students must understand in order to make independent contributions; or
- a summation/expansion of material presented at a recent workshop, symposium or keynote address.

SpringerBriefs in Control, Automation and Robotics allows authors to present their ideas and readers to absorb them with minimal time investment, and are published as part of Springer's e-Book collection, with millions of users worldwide. In addition, Briefs are available for individual print and electronic purchase.

Springer Briefs in a nutshell

- 50 – 125 published pages, including all tables, figures, and references;
- softcover binding;
- publication within 9–12 weeks after acceptance of complete manuscript;
- copyright is retained by author;
- authored titles only – no contributed titles; and
- versions in print, eBook, and MyCopy.

Indexed by Engineering Index.

Publishing Ethics: Researchers should conduct their research from research proposal to publication in line with best practices and codes of conduct of relevant professional bodies and/or national and international regulatory bodies. For more details on individual ethics matters please see: https://www.springer.com/gp/authors-editors/journal-author/journal-author-helpdesk/publishing-ethics/14214

More information about this subseries at http://www.springer.com/series/10198

Xi-Ren Cao

Foundations of Average-Cost Nonhomogeneous Controlled Markov Chains

Springer

Xi-Ren Cao
Department of Automation
Shanghai Jiao Tong University
Shanghai, China

Department of Electronic and Computer Engineering
and Institute of Advanced Study
The Hong Kong University of Science and Technology
Hong Kong, China

ISSN 2191-8112 ISSN 2191-8120 (electronic)
SpringerBriefs in Electrical and Computer Engineering
ISSN 2192-6786 ISSN 2192-6794 (electronic)
SpringerBriefs in Control, Automation and Robotics
ISBN 978-3-030-56677-7 ISBN 978-3-030-56678-4 (eBook)
https://doi.org/10.1007/978-3-030-56678-4

This Springer imprint is published by the registered company Springer Nature Switzerland AG
The registered company address is: Gewerbestrasse 11, 6330 Cham, Switzerland

Preface

Many real-world systems are time-dependent, and therefore, their state evolution should be modeled as time-nonhomogeneous processes. However, performance optimization of time-nonhomogeneous Markov chains (TNHMCs) with the long-run average criterion does not seem well presented in the existing books in the literature.

In a TNHMC, the state spaces, transition probabilities, and reward functions depend on time; and notions such as stationarity, ergodicity, periodicity, connectivity, recurrent and transient states, no longer apply. These properties are crucial to the analysis of time-homogeneous Markov chains (THMCs). Besides, the long-run average suffers from the so-called under-selectivity, which roughly means that it does not depend on the rewards received in any finite period. Dynamic programming may not be very suitable to address the under-selectivity problem because it treats every time instant equally. Therefore, new notions and a different optimization approach are needed.

In this book, we provide comprehensive treatment for the optimization of TNHMCs. We first introduce the notion of confluencity of states, which refers to the property that two independent sample paths of a Markov chain starting from two different initial states will eventually meet together in a finite period. We show that confluencity captures the essentials of the performance optimization of both TNHMCs and THMCs. With confluencity, the states in a TNHMC can be classified into different classes of confluent states and branching states, and single-class and multi-class optimization can be implemented.

We apply the relative optimization approach to the problems. It is simply based on comparing the performance measures of a system under any two policies, and so is also called a *sensitivity-based approach*. The development of this approach started from the early 80s, with the perturbation analysis of discrete-event dynamic systems. The approach is motivated by the study of information technology systems, which usually have large sizes and unknown system structures or parameters. Because of the complexity of the systems, the analysis is usually based on sample paths, so the approach is closely related to reinforcement learning. The approach has been successfully applied to the performance optimization of THMCs.

With confluencity and relative optimization, the problems related to the optimization of TNHMCs can be solved in an intuitive way. State classification is carried out with confluencity; the optimization conditions for the long-run average of single-class and multi-class TNHMCs, bias, *N*th-bias, and Blackwell optimality are derived by the relative optimization approach; and the under-selectivity is reflected in the optimality conditions. The optimization of THMCs becomes a special case.

In summary, the book distinguishes itself by a new notion, the confluencity, a new approach, the relative optimization, and the new results related to TNHMCs developed based on them. The book shows that confluencity captures the essentials of the performance optimization for both THMCs and TNHMCs, and that relative optimization fits TNHMCs better than dynamic programming.

The book is a continuation of two of my previous books, *Stochastic Learning and Optimization—A Sensitivity-Based Approach*, Springer, 2007, and *Relative Optimization of Continuous-Time and Continuous-State Stochastic Systems*, Springer, 2020. The readers may refer to these two books for more about the relative optimization approach and its applications.

Most parts of my works in this book were done when I was with the Institute of Advanced Study, The Hong Kong University of Science and Technology, and the Department of Finance and the Department of Automation, Shanghai Jiao Tong University. I would like to express my sincere appreciation to these two excellent institutes for the nice research environment and financial support. Lastly, I would like to express my gratitude to my wife, Mindy, my son and daughter in law, Ron and Faye, and my two lovely grandsons, Ryan and Zack, for the happiness that they have brought to me, which has made my life colorful.

Hong Kong
June 2020

Xi-Ren Cao
xrcao@sjtu.edu.cn
eecao@ust.hk

Contents

1 Introduction 1
1.1 Problem Formulation 4
1.1.1 Discrete-Time Discrete-State Time-Nonhomogeneous Markov Chains 4
1.1.2 Performance Optimization 5
1.2 Main Concepts and Results 8
1.3 Review of Related Works 10

2 Confluencity and State Classification 13
2.1 The Confluencity 13
2.1.1 Confluencity and Weak Ergodicty 13
2.1.2 Coefficient of Confluencity 16
2.2 State Classification and Decomposition 18
2.2.1 Connectivity of States 18
2.2.2 Confluent States 20
2.2.3 Branching States 23
2.2.4 State Classification and Decomposition Theorem 25

3 Optimization of Average Rewards and Bias: Single Class 29
3.1 Preliminary 29
3.2 Performance Potentials 32
3.2.1 Relative Performance Potentials 32
3.2.2 Performance Potential, Poisson Equation, and Dynkin's Formula 34
3.3 Optimization of Average Rewards 37
3.3.1 The Performance Difference Formula 37
3.3.2 Optimality Condition for Average Rewards 38
3.4 Bias Optimality 44
3.4.1 Problem Formulation 44
3.4.2 Bias Optimality Conditions 48

3.5 Conclusions 53
3.6 Appendix 53

4 Optimization of Average Rewards: Multi-Chains 59
4.1 Performance Potentials of Multi-class TNHMCs 59
4.2 Optimization of Average Rewards: Multi-class 64
4.2.1 The Performance Difference Formula 64
4.2.2 Optimality Condition: Multi-Class 65
4.3 Conclusion 71
4.4 Appendix. The "lim inf" Performance and Asynchronicity 71

5 The *N*th-Bias and Blackwell Optimality 79
5.1 Preliminaries 80
5.2 The *N*th-Bias Optimality 82
5.2.1 The *N*th Bias and Main Properties 82
5.2.2 The *N*th-Bias Difference Formula 86
5.2.3 The *N*th-Bias Optimality Conditions I 89
5.2.4 The *N*th-Bias Optimality Conditions II 93
5.3 Laurent Series, Blackwell Optimality, and Sensitivity Discount Optimality 100
5.3.1 The Laurent Series 100
5.3.2 Blackwell Optimality 102
5.3.3 Sensitivity Discount Optimality 104
5.3.4 On Geometrical Ergodicity 106
5.4 Discussions and Extensions 108

Glossary 109

References 113

Series Editors' Biographies 117

Index 119

Chapter 1
Introduction

This book deals with the performance optimization problem related to the discrete-time discrete-state time-nonhomogeneous Markov chains (TNHMCs), in which the state spaces, transition probabilities, and reward functions, depend on time. There are many excellent books in the literature on performance optimization of the discrete-time discrete-state time-homogeneous Markov chains (THMCs) (see, e.g., Altman (1999), Bertsekas (2007), Cao (2007), Feinberg and Shwartz (2002), Hernández-Lerma and Lasserre (1996, 1999), Puterman (1994)). This book, however, provides comprehensive treatment for the optimization of TNHMCs, in particular, with the long-run average performance.

The analysis of TNHMCs encounters a few major difficulties: (1) Notions such as stationarity, ergodicity, periodicity, connectivity, recurrent and transient states, which are crucial to the analysis of THMCs, no longer apply. (2) The long-run average-reward criterion is under-selective; e.g., the long-run average does not depend on the decisions in any finite period, and thus dynamic programming, which treats every time instant separately and equally, is not very suitable for the problem.

Therefore, we need to formulate new notions and apply a different approach, other than dynamic programming, to the problem. In this book, we introduce the "*confluencity*" of states and show that it is the essential property for performance optimization. Confluencity refers to the property that two independent sample paths of a Markov chain starting from any two different initial states will eventually meet together in a finite period. This notion enables us to develop properties that may replace the ergodicity, stationarity, connectivity, and aperiodicity, etc., used in the optimization of THMCs. With confluencity, we may classify the states into different classes, and develop optimality conditions for the long-run average performance for both uni-chain and multi-class TNHMCs.

Furthermore, we apply the *relative optimization* approach, also called the *direct-comparison-based* or the *sensitivity-based* approach, instead of dynamic programming, to the problem. The approach is based on a formula that gives the difference

X.-R. Cao, *Foundations of Average-Cost Nonhomogeneous Controlled Markov Chains*, SpringerBriefs in Control, Automation and Robotics,
https://doi.org/10.1007/978-3-030-56678-4_1

between the performance measures of any two policies. The formula contains global information about the performance difference in the entire time horizon, and the approach provides a new view to performance optimization and also may solve the issue of under-selectivity.

Although the principles discussed in the book apply to other performance criteria, such as the finite period and discounted total reward, as well, we will focus only on the average-cost related criteria, including the long-run average, bias, Nth-bias, and Blackwell optimality. In economics, a total discounted reward is often used as the performance criterion. However, in computer and communication networks, because of the rapid transition and quick decision-making, cost-discounting does not make sense; for example, there is no point in discounting the waiting time of a packet. As pointed out by Puterman, "*Consequently, the average-reward criterion occupies a cornerstone of queueing control theory especially when applied to control computer systems and communication networks.*" (Puterman 1994).

The principles presented in this book apply to THMCs naturally. Therefore, the state confluencity and the relative optimization indeed form a foundation of the performance optimization of Markov systems, TNHMCs and THMCs.

There are many excellent results on the optimization and control of time-homogeneous systems (see, e.g., Altman (1999), Arapostathis et al. (1993), Åström (1970), Bertsekas (2007), Bryson and Ho (1969), Cao (2007), Feinberg and Shwartz (2002), Hernández-Lerma and Lasserre (1996, 1999), Meyn and Tweedie (2009), Puterman (1994)); every such result should have its counterpart in nonhomogeneous systems. In this sense, the results of this book have wide applications. Here are some simple examples.

Example 1.1 A communication server with fixed-length packets can be modeled by a discrete-time birth-death process (Kleinrock 1975; Sericola 2013). Time is divided into intervals with a fixed length, denoted by $k = 0, 1, \ldots$. In the homogeneous case, a packet arrives with probability $1 < \lambda < 1$ at every time k, and the server transmits a packet in a time slot with probability $0 < \mu(n) < 1$, where n is the number of packets in the server. The server has a buffer with a maximum size of N, i.e., if the number of packets is N, then any arriving packet will be rejected and lost; so $0 \leq n \leq N$. The arrival and service processes are independent. Let $p_n := \lambda[1 - \mu(n)]$, $q_n := \mu(n)[1 - \lambda]$, $0 < n < N$, $\mu(0) = 0$, $p_N = 0$, and $q_N = \mu(N)$; and p_n are the birth rate and q_n are the death rate, $0 \leq n \leq N$. The number of packets in the server can be modeled with a discrete birth-death process $\boldsymbol{X} = \{X_0, X_1, \ldots\}$, $X_k \in \mathscr{S} := \{0, 1, \ldots, N\}$, with the following transition probabilities: for $0 < n < N$,

$$P(X_{k+1}|X_k) = \begin{cases} p_n & if\ X_k = n,\ X_{k+1} = n + 1, \\ q_n & if\ X_k = n,\ X_{k+1} = n - 1, \\ 1 - (p_n + q_n) & if\ X_k = X_{k+1} = n; \end{cases}$$

and $P(N - 1|N) = q_N$, $P(N|N) = 1 - q_N$, $P(1|0) = \lambda$, and $P(0|0) = 1 - \lambda$.

With the control policy $\mu(n)$, the performance criterion to be optimized is the long-run average (discounting makes no sense due to rapid transmissions)

$$\eta := \lim_{K\to\infty} \left\{ \frac{1}{K} E\Big[\sum_{k=0}^{K-1} f(X_k) \Big| X_0 = 0 \Big] \right\},$$

with the reward function $f(n) = n$, for $0 < n < N$, and $f(N) = -\frac{N}{2}$, which represents a penalty for losing packets when $X_k = N$. Under some mild conditions, $\boldsymbol{X}$ is ergodic, and the problem can be solved by the standard Markov decision process (MDP) approach in the literature, e.g., Hernández-Lerma and Lasserre (1996), Puterman (1994).

In general, the arrival rate λ_k may depend on time k, so do the transition probabilities $P_k(X_{k+1}|X_k)$, representing the time-dependent nature of the demand. The Markov chain $\boldsymbol{X}$ is no longer ergodic or stationary, and $\boldsymbol{X}$ may wander around the state space $\mathscr{S}$ forever and never settles down (in distribution). Then, the question is, what can we do for these nonhomogeneous systems? This question is addressed in general in the rest of the book. □

Example 1.2 Consider a one-dimensional linear quadratic Gaussian (LQG) control problem (Bertsekas 2007; Bryson and Ho 1969; Cao 2007). The system dynamic is described by a linear equation. In the time-homogeneous case, it is

$$X_{k+1} = aX_k + bu_k + w, \tag{1.1}$$

where $X_k \in \mathscr{R}$ is the system state, u_k is the control variable, w is a Gaussian distributed random noise, at time k, and a and b are constants. The reward function is in a quadratic form:

$$\eta = \lim_{K\to\infty} \left\{ \frac{1}{K} \sum_{k=0}^{K-1} E[(qX_k^2 + ru_k^2)|X_0 = x] \right\}, \tag{1.2}$$

It is proved that a linear feedback control $u_k = -dX_k$, $k = 0, 1, \ldots$, achieves the optimal performance (Bryson and Ho 1969; Cao 2007). Under this linear policy, the system equation (1.1) becomes $X_{k+1} = cX_k + w$, with $c = a - bd$. The system is stable with $c < 1$. This problem can be easily solved within the standard time-homogeneous MDP framework.

In the time-nonhomogeneous case, the system parameters are time dependent: $X_{k+1} = a_k X_k + b_k u_k + w_k$, the reward function becomes $f_k(x, u) = q_k x^2 + r_k u^2$. Because a_k and b_k are different at different $k's$, the system may move arbitrarily on $\mathscr{R}$. The Markov chain $\boldsymbol{X}$ may not reach a stationary status, and the performance defined in (1.2) may not exists. Then, the question is, how can we analyze such nonhomogeneous systems in a general way? □

The simple model of LQG can be used in the more advanced study in the *mean-field game* theory (Caines 2019; Bensoussan et al. 2013), which was originally motivated by a power control problem in a wireless communication network. Other applications include the study of the transient behavior of any physical system; for example, in launching a satellite, it will go around the earth a few times before reaching its final orbit; so the system is nonhomogeneous before it reaches the stationary status (see Chap. 5). Time-nonhomogeneous Markov models are also widely used in other fields such as economics and social networks (Bhawsar et al. 2014; Briggs and Sculpher 1998; Proskurnikov and Tempo 2017).

1.1 Problem Formulation

1.1.1 *Discrete-Time Discrete-State Time-Nonhomogeneous Markov Chains*

Consider a discrete-time discrete-state TNHMC $\boldsymbol{X} := \{X_k, X_k \in \mathscr{S}, k = 0, 1, \ldots\}$, where $\mathscr{S} = \{1, 2, \ldots, S\}$ denotes the *state space*, the space of the states at all time instants $k = 0, 1, \ldots$, with $S = |\mathscr{S}|$ being the number of states in $\mathscr{S}$ (S may be infinity).

A time-nonhomogeneous Markov chain may stroll around in the state space, which means that at different times k, $k = 0, 1, \ldots$, the Markov chain may be at different subspaces of $\mathscr{S}$. We allow new states to join the chain at any time $k = 0, 1, \ldots$. Let $\mathscr{S}_k \subseteq \mathscr{S}$ be the *state space* at time $k = 0, 1, \ldots$. It includes the states joining the chain at k. Let $P_k(y|x)$ denote the transition probability from state x to state y at time k; then, we have

$$\mathscr{S}_k = \{y \in \mathscr{S} : P_k(z|y) > 0, \ for\ some\ z \in \mathscr{S}\}, \tag{1.3}$$

and it is also called the *input set* of the Markov chain at k, $k = 0, 1, \ldots$; and we further define

$$\mathscr{S}_{k,out} = \{y \in \mathscr{S} : P_k(y|z) > 0 \ for\ some\ z \in \mathscr{S}_k\} \tag{1.4}$$

as the *output set* of the chain at k, $k = 0, 1, \ldots$. For the Markov chain to run properly, it requires that

$$\mathscr{S}_{k,out} \subseteq \mathscr{S}_{k+1}, \quad k = 0, 1, \ldots. \tag{1.5}$$

When $\mathscr{S}_{k,out} \subset \mathscr{S}_{k+1}$, there are states joining the chain at time $k+1$. Let $S_k := |\mathscr{S}_k|$ be the number of the states in the state space, or the input set, at k, $k = 0, 1, \ldots$. We set $P_k := [P_k(z|y)]_{y \in \mathscr{S}_k, z \in \mathscr{S}_{k+1}}$ to be the *transition probability matrix* at k and it may not be a square matrix, and set $\mathbb{P} = \{P_0, P_1, \ldots P_k, \ldots\}$, $\mathbb{S} = \{\mathscr{S}_0, \mathscr{S}_1, \ldots, \mathscr{S}_k, \ldots\}$. $\mathbb{P}$ is called a *transition law* of the process $\boldsymbol{X}$.

An example of possible state spaces and transitions are shown in Fig. 1.1. In the figure, $\mathscr{S}_{2,out} = \{x_4, x_5, x_7\} \subset \mathscr{S}_3 = \{x_4, x_5, x_6, x_7\}$; states x_6 joins the chain at time 3. Similarly, states x_3 joins the chain at time 2. The states at different times

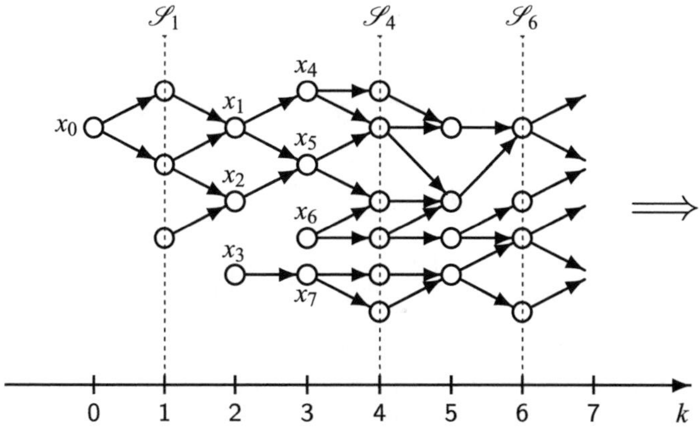

Fig. 1.1 State spaces and state transitions of a TNHMC: circles represent states, arrows represent transitions, and state spaces $\mathscr{S}_1$, $\mathscr{S}_4$, and $\mathscr{S}_6$ are indicated

may be completely different, although they are put on the same horizontal line for convenience.

Example 1.3 All the time-homogeneous Markov chains, including irreducible chains, uni-chains, or multi-chains, are special cases of time-nonhomogeneous Markov chains with $P_k \equiv P$ and $\mathscr{S}_k = \mathscr{S}$ for all $k = 0, 1, \ldots$. □

1.1.2 Performance Optimization

Let $(\Omega, \Sigma, \mathscr{P})$ denote the probability space of the Markov chain generated by the transition law $\mathbb{P}$ and the initial probability measure of X_0; sometimes, we may denote this probability measure by $\mathscr{P}^{\mathbb{P}} = \mathscr{P}$. Let $E^{\mathbb{P}} = E$ be the corresponding expectation. We have $\mathscr{P}(X_k = y|X_0 = x) > 0$ for some $y \in \mathscr{S}_{k-1,out}$ and $x \in \mathscr{S}_0$ (we use the condition $\{\bullet|X_0 = x\}$ to denote the initial state).

First, we denote the reward function at time k by $f_k(x)$, $x \in \mathscr{S}_k$, $k = 0, 1, \ldots$, and we denote the reward vector at time k by $f_k = (f_k(1), \ldots, f_k(i), \ldots, f_k(S_k))^T$, and also set $\mathscr{f} = \{f_0, f_1, \ldots, f_k, \ldots\}$.

The theory for finite-horizon performance optimization problems for TNHMCs is almost the same as that for the THMCs because the finite-sum performance measure $E\{\sum_{l=k}^{K} f(X_l)|X_k = x\}$ also depends on time k for THMCs. Therefore, in this book, we focus on infinite horizon problems. The long-run average reward, starting from $X_k = x$, is defined by

$$\eta_k(x) = \liminf_{K\to\infty} \frac{1}{K} E\Big\{ \sum_{l=k}^{k+K-1} f_l(X_l) \Big| X_k = x \Big\}. \tag{1.6}$$

If the limit in (1.6) exists, then we have

$$\eta_k(x) = \lim_{K\to\infty} \frac{1}{K} E\Big\{ \sum_{l=k}^{k+K-1} f_l(X_l) \Big| X_k = x \Big\}. \tag{1.7}$$

The discounted performance is defined by

$$\upsilon_{\beta,k}(x) := E\left\{ \sum_{l=0}^{\infty} \beta^l f_{k+l}(X_{k+l}) \Big| X_k = x \right\}, \tag{1.8}$$

$0 < \beta < 1$ is called a *discount factor*. In (1.6)–(1.8), $\eta_k(x)$ or $\eta_{\beta,k}(x)$ is called a *performance measure*, or a *performance criterion*.

While the long-run average measures the infinite horizon behavior, there is another performance measure called "bias", which measures the transient behavior of the system. For bias, we assume that the limit (1.7) exists; and the bias is defined by

$$g_k(x) = E\Big\{ \sum_{l=k}^{\infty} [f_l(X_l) - \eta_l(X_l)] \Big| X_k = x \Big\}. \tag{1.9}$$

Other performance measures such as the Nth biases, $N = 1, 2, \ldots$, and the Blackwell optimality, will be discussed in the book, and they will be precisely defined in Chap. 5. These definitions essentially take the same forms as those for THMCs in the literature (Altman 1999; Cao 2007; Jasso-Fuentes and Hernández-Lerma 2009a, b; Feinberg and Shwartz 2002; Hernández-Lerma and Lasserre 1996, 1999; Hordijk and Yushkevich 1999a, b; Lewis and Puterman 2001, 2002; Puterman 1994).

To simplify the discussion, we make the following assumption:

Assumption 1.1 (a) The sizes of state spaces $\mathscr{S}_k$ are bounded, i.e., $S_k := |\mathscr{S}_k| \le M_1 < \infty$ for all $k = 0, 1, \ldots$; and

(b) The reward functions are bounded, i.e.,

$$|f_k(x)| < M_2 < \infty. \tag{1.10}$$

Under this assumption, $\eta_k(x)$ is finite for any $x \in \mathscr{S}_k$.

In an optimization problem, at time $k = 0, 1, \ldots$, with $X_k = x \in \mathscr{S}_k$, we may take an action $\alpha_k(x)$ chosen in an action set $\mathscr{A}_k(x)$; the action determines the transition probabilities at x, denoted as $P_k^{\alpha_k(x)}(y|x)$, $y \in \mathscr{S}_{k,out}$, and the reward at state x, denoted by $f_k^{\alpha_k(x)}(x)$. The mapping, or the vector, $\alpha_k = \alpha_k(x)$, $x \in \mathscr{S}_k$, is called a *decision rule*, and the space of all the decision rules at time k can be denoted by $\mathscr{A}_k := \prod_{x\in\mathscr{S}_k} \mathscr{A}_k(x)$, where $\prod$ denotes the Cartesian product. A decision rule at k determines a transition probability matrix at k, $P_k^{\alpha_k} = [P_k^{\alpha_k(x)}(y|x)]_{x\in\mathscr{S}_k, y\in\mathscr{S}_{k,out}}$, $k = 0, 1, \ldots$, and a reward function (or vector) $f_k^{\alpha_k} = (f_k^{\alpha_k(1)}(1), \ldots, f_k^{\alpha_k(S_k)}(S_k))^T$.

A *policy* is a sequence of decision rules at $k = 0, 1, \ldots$, $u := \{\alpha_0, \alpha_1, \ldots\}$, which determines $\mathbb{P} := \{P_0^{\alpha_0}, P_1^{\alpha_1}, \ldots\}$ and $\mathscr{f} := \{f_0^{\alpha_0}, f_1^{\alpha_1}, \ldots\}$. We also refer to a pair $(\mathbb{P}, \mathscr{f}) =: u$ as a policy. As in the case of the standard Markov decision process, we assume that the actions at different states and different $k = 0, 1, \ldots$, can be chosen independently; thus, in every P_k, $k = 1, 2, \ldots$, we may independently choose every row. We do not consider problems with constraints in which an action at state $X_k = x$ may affect the action chosen at any other state $X_{k'} = y$, $k' = \text{or} \neq k$.

A Markov chain and its reward are determined by a policy $u = (\mathbb{P}, \mathscr{f})$ that is applied to the chain. We use a superscript u, or simply $\mathbb{P}$ when $\mathscr{f}$ does not take a role, to denote the quantities associated with the Markov chain controlled by policy u. Thus, the Markov chain is denoted by $\boldsymbol{X}^u$, the input and output sets at time k are denoted by $\mathscr{S}_k^u$ and $\mathscr{S}_{k,out}^u$, $k = 0, 1, \ldots$, respectively, and the decision rule is denoted by α_k^u; and the average reward (1.6) is now denoted by

$$\eta_k^u(x) = \liminf_{K \to \infty} \frac{1}{K} E^u \Big\{ \sum_{l=k}^{k+K-1} f_l^{\alpha_l^u}(X_l^u) \Big| X_k^u = x \Big\}.$$

Let $\mathscr{D}$ denote the space of all admissible policies (to be precisely defined later in Definitions 3.2, 4.1, and 5.1). To make the performance of all policies in $\mathscr{D}$ comparable, we require that the average rewards starting from all states are well defined under all policies. So we need the following assumption (cf. (1.5)).

Assumption 1.2 The state spaces at any k, $k = 0, 1, \ldots$, are the same for all policies; i.e., for any $u, u' \in \mathscr{D}$, we have $\mathscr{S}_k^u = \mathscr{S}_k^{u'} =: \mathscr{S}_k$, $k = 0, 1 \ldots$. □

This assumption is very natural. As shown in Fig. 1.1, new states may be added to the state space; thus, the assumption does not mean that all policies have the same output space; it only requires that every policy assigns transition probabilities to all possible states in $\mathscr{S}_k$. In other words, no state should stop the Markov chain from running under any policy, so its performance does exist. In a THMC, this simply says that all policies have the same state space.

The goal of performance optimization is to find a policy in $\mathscr{D}$ that maximizes the performance criterion (if the maximum can be achieved); i.e., to identify an optimal policy denoted by u^*:

$$u^* = \arg\Big\{ \max_{u \in \mathscr{D}} (\eta_k^u(x)) \Big\}, \quad x \in \mathscr{S}_k, \ k = 0, 1, \ldots. \tag{1.11}$$

If the maximum cannot be attained, the goal is to find the value

$$\eta_k^*(x) = \sup_{u \in \mathscr{D}} (\eta_k^u(x)), \quad x \in \mathscr{S}_k, \ k = 0, 1, \ldots. \tag{1.12}$$

Similar definitions apply to the bias and Nth-bias optimality as well; see the related chapters.

1.2 Main Concepts and Results

In discrete-time and discrete-state TNHMCs, the state spaces, transition probability matrices, and reward functions at different time $k = 0, 1, \ldots$ may be different. As such, notions such as stationarity, ergodicity, periodicity, connectivity, recurrent and transient states, which are crucial to the analysis of THMCs, no longer apply.

Besides, the long-run average performance criterion possesses the *under-selectivity*, which roughly refers to the property that the long-run average does not depend on the actions in any finite period, and thus the optimality conditions do not need to hold in any finite period. Under-selectivity makes dynamic programming not suitable for the problem, because the dynamic programming principle treats every time instant equally, resulting in the same conditions at all time instants.

Therefore, to carry out the performance optimization of TNHMCs, we need to formulate new notions capturing the main features mentioned above. We also need to apply a different approach other than dynamic programming. Optimality conditions, necessary and sufficient, for long-run average, bias, Nth biases, and Blackwell, etc., for single or multi-class TNHMCs, can then be derived.

The main results of the book are summarized as follows. Many of them have not been covered by the books in the literature; this makes this book unique.

(1) **(Confluencity)** We discover that the crucial property in the performance optimization, for both TNHMCs and THMCs, is the *confluencity*. Two states are confluent to each other, if two independent sample paths of the Markov chain starting from these two states, respectively, eventually meet at the same state with probability one. We show that this notion plays a fundamental role in the performance optimization and state classification, and optimization theory can be developed without ergodicity, stationarity, connectivity, and aperiodicity, etc. We also show that confluencity implies weak ergodicity, and explains their differences and why the former is needed for performance optimization and state classification of TNHMCs. These results are discussed in Sect. 2.1.

(2) **(Relative optimization)** We apply the *relative optimization* approach, which has been successfully applied to many optimization problems, see Cao (2020b, 2007), Guo and Hernández-Lerma (2009) and the references therein. The approach is based on a performance difference formula that compares the performance measures of any two policies. While dynamic programming provides only "local" information at any particular time and state, the performance difference formula contains the "global" information about the performance comparison in the entire horizon (see the discussion in Chap. 1 of Cao 2020a). The approach is also called the *sensitivity-based* or *direct-comparison*-based approach in the literature (Cao 2007).

(3) **(Uni-chain optimization with under-selectivity)** A TNHMC is said to be a uni-chain if all the states in the chain are confluent to each other. We define the *performance potential* for uni-chains, which serves as a building block of performance optimization, and it extends the same notion of potentials of THMCs (Cao 2007). With performance potentials, we may derive the performance differ-

ence formula, and then obtain the necessary and sufficient optimality conditions for the long-run average rewards. From the difference formula, it is clear that the optimality conditions may not need to hold in any finite-time period, so the under-selectivity is reflected in these conditions. These results extend the well-known optimization results for THMCs, see, e.g., Altman (1999), Arapostathis et al. (1993), Bertsekas (2007), Feinberg and Shwartz (2002), Hernández-Lerma and Lasserre (1996, 1999), Puterman (1994); they are discussed in Chap. 3.

(4) **(State classification)** We show that with confluencity, all the states in a TNHMC can be classified into confluent states and branching states, and all the confluent states can be grouped into a number of confluent classes. Independent sample paths starting from the same confluent state meet together infinitely often with probability one (w.p.1); the sample paths starting from different confluent states in the same class, even at different times, will meet w.p.1 as well; the sample paths from states in different confluent classes, however, will never meet; and a sample path starting from a branching state will enter one of at least two confluent classes and stay there forever. Confluent and branching are extensions of (but somewhat different from) the notions of recurrent and transient (Puterman 1994), and the former is a better state classification than the latter, in the sense that the confluent and branching are more appropriate to capture the relations of the states in performance optimization. These results are discussed in Sect. 2.2.

(5) **(Multi-class optimization)** We derive both necessary and sufficient conditions for the optimal policies of the long-run average reward of multi-chain TNHMCs (consisting of multiple confluent classes and branching states). Surprisingly, these results look very similar to those for THMCs (Cao 2007; Guo and Hernández-Lerma 2009; Puterman 1994). These results also reflect the under-selectivity; e.g., the conditions do not need to hold for any finite period. These results are discussed in Chap. 4.

(6) **(Bias optimality)** The long-run average only measures the asymptotic behavior, and it does not depend on the behavior in the initial period. The most important measure of transient performance is the *bias* defined by (1.9). Bias optimality refers to the optimization of bias in the space of all optimal long-run average policies. This problem has been well-studied for THMCs in the literature (Guo et al. 2009; Hernández-Lerma and Lasserre 1999; Jasso-Fuentes and Hernández-Lerma 2009b; Lewis and Puterman 2000, 2001, 2002). With confluencity and relative optimization, we extend the results to TNHMCs. Interestingly, under-selectivity is also reflected in the bias optimization conditions. These results are discussed in Sect. 3.4.

(7) **(N th-bias optimality and Blackwell optimality)** While the bias measures transient performance, we observe that a bias has a bias, which has a bias too, and so on. In general, the $(N+1)$th bias is the bias of the Nth bias, $N \geq 1$ (with the first bias being the bias in (1.9)). The Nth biases, $N = 1, 2, \ldots$, describe the transient behavior of a Markov chain at different levels. The policy that optimizes all the Nth biases, $N = 0, 1, \ldots$, is the one with no under-selectivity. A Blackwell optimal policy is optimal for discounted performance measures (1.8) for all discount factors near one. The discounted reward (1.8) can be expressed as a Laurent

series in all the Nth biases, and the policy optimizing all the Nth biases is the Blackwell optimal policy and vice versa. It is shown that the Nth-bias approach is equivalent to the sensitive discount optimality; These results are well established for THMCs, see Cao (2007), Guo and Hernández-Lerma (2009), Hordijk and Yushkevich (1999a, b), Jasso-Fuentes and Hernández-Lerma (2009a), Puterman (1994), Veinott (1969), and in this book, we extend them to TNHMCs. The necessary and sufficient optimality conditions for the Nth-bias optimal and Blackwell optimal policies are derived. These results are discussed in Chap. 5.

The Nth-bias optimality provides a complete solution to the under-selectivity of long-run average performance. The approach does not resort to discounting.

All the optimality results are obtained by the relative optimization approach, based on a direct comparison of the performance measures, or the biases, or Nth biases, under any two policies. They demonstrate the applications of relative optimization.

1.3 Review of Related Works

Many research works have been done regarding the issues discussed in this book, dating back to Kolmogoroff (Kolmogoroff 1936) and Blackwell (Blackwell 1945). Main results include

(1) Concepts similar to ergodicity and stationarity, called *weak ergodicity* and *stability*, were proposed. Weak ergodicity, or merging, refers to the property that a Markov chain asymptotically forgets where it started (i.e., as time passes by, the state probability distribution becomes asymptotically independent of the initial state (Hajnal 1958; Park et al. 1993; Saloff-Coste and Zuniga 2010, 2007). Stability relates to whether the state probability distributions will converge, or become close to each other, in the long term (Meyn and Tweedie 2009; Saloff-Coste and Zuniga 2010). A technique called *coupling* is closely related to the confluencity, but both have different connotations, and they are introduced for different purposes. Coupling consists of two copies of a Markov chain running simultaneously; they behave exactly like the original Markov chain but may be correlated. It was used widely in studying stationarity, weak ergodicity, and the rate of convergence, and many other problems (Cao 2007; Doeblin 1937; Griffeath 1975a, b, 1976; Pitman 1976; Roberts and Rosenthal 2004).

(2) The decomposition–separation (DS) theorem (Blackwell 1945; Cohn 1989; Kolmogoroff 1936; Sonin 1991, 1996, 2008) was established, which claims that, with probability one, any sample path of a TNHMC will, after a finite number of steps, enter one of the "jets" (a sequence of subsets of the state spaces at time $k = 0, 1, \ldots$). The DS theorem can be viewed as a decomposition theorem of sample paths, which is closely related to the state decomposition for TNHMCs introduced in this book. In Platis et al. (1998), the computation of the hitting time from a state to a subset of states in a time-nonhomogeneous chain was

discussed, and sufficient conditions for the existence of the mean hitting time were provided.

(3) The under-selectivity of average rewards of TNHMCs were discussed, e.g., in Bean and Smith (1990), Hopp et al. (1987), Park et al. (1993). To take the finite-period performance into consideration, Hopp et al. (1987) introduced the *periodic forecast horizon (PFH) optimality*, and proved that under some conditions, PFH optimality implies average optimality. This roughly means that the PFH optimal policy, as an average optimal policy, is the limit point, defined in some metric, of a sequence of optimal policies for finite horizon problems. Although not mentioned in Hopp et al. (1987), their results apply only to the "uni-chain" case defined in this book.

(4) There are many excellent works on control and optimization of THMCs, and we cannot cite all of them in this book; they include (Altman 1999; Bertsekas 2007; Cao 2007; Feinberg and Shwartz 2002; Hernández-Lerma and Lasserre 1996, 1999; Puterman 1994), to name a few. In addition, the optimization of total rewards for TNHMCs was discussed in Hinderer (1970), and the results are similar to the homogeneous chains (Hernández-Lerma and Lasserre 1996; Puterman 1994).

(5) The Nth-bias optimization and its relation with the Blackwell optimality were presented in Cao (2007), Cao and Zhang (2008b) for discrete-time discrete-state THMCs, and in Zhang and Cao (2009), Guo and Hernández-Lerma (2009) for continuous-time discrete-state multi-chain THMCs. It is equivalent to the sensitive discount optimality, which was first proposed by Veinott in Veinott (1969), and has been studied by many researchers, see, e.g., Dekker and Hordijk (1988), Guo and Hernández-Lerma (2009), Hordijk and Yushkevich (1999a), Hordijk and Yushkevich (1999b), Jasso-Fuentes and Hernández-Lerma (2009a), Puterman (1994). Also, there are many works on bias optimality for THMCs, e.g., Guo et al. (2009), Lewis and Puterman (2001), Puterman (1994), Jasso-Fuentes and Hernández-Lerma (2009a), Jasso-Fuentes and Hernández-Lerma (2009b).

(6) Most of the contents of this book are from the authors' research works reported in Cao (2019a, 2016, 2015). In Cao (2015), the notion of confluencity is proposed, and the necessary and sufficient optimality conditions for the long-run average and bias are derived for uni-chain TNHMCs. In Cao (2016), confluencity is used to classify the states, and the necessary and sufficient optimality conditions for the long-run average are derived for multi-class TNHMCs. In Cao (2019a), the Nth-bias and Blackwell optimality are discussed with no discounting.

(7) The development of relative optimization started with the perturbation analysis of discrete-event dynamic systems (Cao 2000; Cao and Chen 1997; Cassandras and Lafortune 2008; Fu and Hu 1997; Glasserman 1991; Ho and Cao 1991). The approach is motivated by the study of information technology systems and is closely related to reinforcement learning (Cao 2007). Its applications to the optimization of THMCs are summarized in Cao (2007). This approach is based on first principles; it starts with a simple comparison of the performance measures of any two policies. It provides global information on the entire horizon and thus may be applied to problems to which dynamic programming may not fit

very well, It has been successfully applied to many problems, e.g., event-based optimization (Cao and Zhang 2008a; Xia et al. 2014), and partially observable Markov decision problems with separation principle extended (Cao et al. 2014), and stochastic control of diffusion processes (Cao 2020a, 2017), to name a few.

(8) There are many other excellent research works on TNHMCs in the current literature. For example, Benaïm et al. (2017), Madsen and Isaacson (1973) discussed the ergodicity and strongly ergodic behavior of TNHMCs, Douc et al. (2004) studied the bounds for convergence rates of TNHMCs; Connors and Kumar (1989, 1988) worked on the recurrence order of TNHMCs and its application to simulated annealing; Cohn (1989, 1972, 1976, 1982) considered a number of important topics related to TNHMCs, including their asymptotic behavior and the limit theorems; Zheng et al. (2018) showed that when the transition probabilities of a TNHMC change slowly over time, various expectations associated with it can be rigorously approximated by means of THMCs; Teodorescu (1980) developed some tools for deriving TNHMCs with desired properties, starting from observed data in real systems; Brémaud (1999) discussed eigenvalues associated with TNHMCs; and Alfa and Margolius (2008), Li (2010), Margolius (2008) studied the transient behavior of TNHMCs with periodic or block structures, with the RG factorization approach; and many more.

Chapter 2
Confluencity and State Classification

In this chapter, we introduce the notion of confluencity and show that with confluencity, the states of a TNHMC can be classified into different classes of confluent states and branching states (Cao 2019, 2016, 2015). Confluent and branching are similar to, but different from, recurrent and transient in THMCs (Cao 2007; Hordijk and Lasserre 1994; Puterman 1994). A branching state is transient, and a recurrent state is confluent, but a transient state may be either a branching or a confluent state, and a confluent state may be either recurrent or transient (see Table 2.1). We will show that the classification of the confluent and branching states is more essential to optimization than that of the recurrent and transient states.

The results of this chapter form a foundation for performance optimization, which will be discussed in detail in the remaining chapters.

2.1 The Confluencity

In this section, we introduce the confluencity of states, discuss its relations with weak ergodicity, and define a "coefficient of confluencity" as a criterion for confluencity.

2.1.1 *Confluencity and Weak Ergodicty*

2.1.1.1 The Confluencity

Consider two independent sample paths starting from time k, $\boldsymbol{X} := \{X_l, l \geq k\}$ and $\boldsymbol{X}' := \{X'_l, l \geq k\}$, generated by the same transition laws $\{P_l, l = k, k+1, \ldots\}$, but

X.-R. Cao, *Foundations of Average-Cost Nonhomogeneous Controlled Markov Chains*, SpringerBriefs in Control, Automation and Robotics,
https://doi.org/10.1007/978-3-030-56678-4_2

with two different initial states $X_k = x, X'_k = y, x, y \in \mathscr{S}_k$. For $k = 0, 1, \ldots$, define the *confluent time* of x and y at time k by

$$\tau_k(x, y) := \min\{\tau \geq k, X_\tau = X'_\tau\} - k, \quad x, y \in \mathscr{S}_k. \tag{2.1}$$

$\tau_k(x, y) \geq 0$ is the time (a random variable) required for the two paths to first meet after k. By definition, $X_{\tau_k(x,y)} = X'_{\tau_k(x,y)}$.

The underlying probability space generated by the two independent sample paths can be denoted by $\Omega \times \Omega$ with a probability measure being an extension of $\mathscr{P}$ on Ω to $\Omega \times \Omega$. For simplicity and without much confusion, we also denote this extended measure on $\Omega \times \Omega$ by $\mathscr{P}$ and its corresponding expectation by "E".

Definition 2.1 A TNHMC $\boldsymbol{X} := \{X_k, k \geq 0\}$ is said to be *confluent*, or to satisfy the *confluencity*, if for all $x, y \in \mathscr{S}_k$ and any $k = 1, 2, \ldots$, $\tau_k(x, y)$ are finite ($< \infty$) with probability one (w.p.1); i.e.,

$$\lim_{K \to \infty} \mathscr{P}(\tau_k(x, y) > K | X_k = x, X'_k = y) = 0. \tag{2.2}$$

A TNHMC is said to be *strongly confluent*, or to satisfy the *strong confluencity*, if

$$E\{\tau_k(x, y) | X_k = x, X'_k = y\} < M_3 < \infty, \tag{2.3}$$

for all $x, y \in \mathscr{S}_k$ and $k = 0, 1, \ldots$. We also say that the two states x and y in (2.2) are confluent to each other. □

Confluencity of a Markov chain implies that the states in $\mathscr{S}_k$, $k = 0, 1, \ldots$, are in some sense connected through future states. A confluent TNHMC corresponds to a uni-chain in THMCs (Puterman 1994). When confluencity does not hold for all states, the states of the Markov chain may be decomposed into multiple confluent classes of states and branching states, the states in each confluent class are confluent to each other; these topics will be discussed in Sect. 2.2.

We coin the word "confluencity" because other synonyms such as "merging" and "coupling" have been used for other connotations. The notion of confluencity is similar to that of "coupling" in the literature (Doeblin 1937; Griffeath 1975a, b; Pitman 1976), but both have different connotations. Coupling refers to the construction of a bivariate process $(\boldsymbol{X}^{(1)}, \boldsymbol{X}^{(2)})$ on the state space $\mathscr{S} \times \mathscr{S}$ in the following way: both $\boldsymbol{X}^{(1)}$ and $\boldsymbol{X}^{(2)}$ have the same probability measure $\mathscr{P}$, but the two processes $\boldsymbol{X}^{(1)}$ and $\boldsymbol{X}^{(2)}$ can be correlated; and if at some time k, $X_k^{(1)} = X_k^{(2)}$, then after k, $\boldsymbol{X}^{(1)}$ and $\boldsymbol{X}^{(2)}$ remain the same forever. Coupling is a *technique* mainly used in studying weak ergodicity and the rate of convergence of transition probabilities (Cao 2007; Doeblin 1937; Griffeath 1975a, b, 1976; Pitman 1976; Roberts and Rosenthal 2004). In Cao (2007), coupling is used to reduce the variance in estimating the performance potentials. Confluencity, on the other hand, refers to a particular *property* of a TNHMC where the two sample paths run independently in a natural way, and no construction is involved; it is used in state classification and performance optimization, as shown in this book.

2.1.1.2 Weak Ergodicity

The confluencity defined in (2.2) plays a similar role for TNHMCs as the connectivity and ergodicity for THMCs. Indeed, (2.2) assures the weak ergodicity, which means that as k goes to infinity, the system will forget the effect of the initial state, as defined below:

Definition 2.2 A TNHMC is said to be *weakly ergodic*, if for any $x, x' \in \mathscr{S}_k$ and any $k, k' = 0, 1, \ldots$, it holds that

$$\lim_{K\to\infty} \left\{ \max_{y\in\mathscr{S}_K} \left| \mathscr{P}(X_K = y|X_k = x) - \mathscr{P}(X_K = y|X_{k'} = x') \right| \right\} = 0. \tag{2.4}$$

Lemma 2.1 *Under Assumption 1.1a, confluencity implies weak ergodicity.*

Proof First, let $k = k'$. By the strong Markov property, for any $x, x' \in \mathscr{S}_k$, we have

$$\begin{aligned}&\mathscr{P}(X_K = y|X_k = x, \tau_k(x, x') < K - k)\\ &= \mathscr{P}(X_K = y|X_k = x', \tau_k(x, x') < K - k), \ \forall y \in \mathscr{S}_K.\end{aligned}$$

Furthermore,

$$\begin{aligned}&\mathscr{P}(X_K = y|X_k = x)\\ &= \mathscr{P}(X_K = y|X_k = x, \tau_k(x, x') < K - k)\mathscr{P}(\tau_k(x, x') < K - k)\\ &\quad + \mathscr{P}(X_K = y|X_k = x, \tau_k(x, x') > K - k)\mathscr{P}(\tau_k(x, x') > K - k).\end{aligned}$$

Therefore,

$$\begin{aligned}&|\mathscr{P}(X_K = y|X_k = x) - \mathscr{P}(X_K = y|X_k = x')|\\ &= |\mathscr{P}(X_K = y|X_k = x, \tau_k(x, x') > K - k)\\ &\quad - \mathscr{P}(X_K = y|X_k = x', \tau_k(x, x') > K - k)|\\ &\quad \times \mathscr{P}(\tau_k(x, x') > K - k)\\ &\le 2\mathscr{P}(\tau_k(x, x') > K - k), \quad \forall y \in \mathscr{S}_K.\end{aligned} \tag{2.5}$$

Then, for $k = k'$, (2.4) follows from the confluencity condition (2.2).

Next, let $k' < k$. We have

$$\begin{aligned}&|\mathscr{P}(X_K = y|X_k = x) - \mathscr{P}(X_K = y|X_{k'} = x')|\\ &= \left|\mathscr{P}(X_K = y|X_k = x) - \sum_{z\in\mathscr{S}} \left\{\mathscr{P}(X_K = y|X_k = z)\mathscr{P}(X_k = z|X_{k'} = x')\right\}\right|\\ &\le \sum_{z\in\mathscr{S}_k} \left\{\left|\mathscr{P}(X_K = y|X_k = x) - \mathscr{P}(X_K = y|X_k = z)\right|\mathscr{P}(X_k = z|X_{k'} = x')\right\}.\end{aligned}$$

Then, for $k' < k$, (2.4) follows from (2.5) and the boundedness of $S_k = |\mathscr{S}_k|$. □

2.1.2 *Coefficient of Confluencity*

Now, we define a *coefficient of confluencity* by

$$\nu := 1 - \inf_k \Big(\min_{x,x' \in \mathscr{S}_k} \{ \sum_{y \in \mathscr{S}_{k,out}} [P_k(y|x) P_k(y|x')] \} \Big).$$

Lemma 2.2 *If* $\nu < 1$, *then the strong confluencity (2.3) holds.*

Proof First, let $\phi_k(x, x') := \sum_{y \in \mathscr{S}_{k,out}} [P_k(y|x) P_k(y|x')]$; it is the probability that two sample paths with $X_k = x$ and $X'_k = x'$ meet at time $k + 1$. Thus, the probability that two sample paths meet at $k + 1$ is at least $\Big[\min_{x,x' \in \mathscr{S}_k} \{\phi_k(x, x')\} \Big]$, and the probability that any two sample paths do not meet at any time k is at most $\nu < 1$. Thus, $\mathscr{P}(\tau_k(x, y) \geq i | X_k = x, X'_k = y) < \nu^{i-1}$, and

$$E\{\tau_k(x, y) | X_k = x, X'_k = y\}$$
$$= \sum_{i=1}^{\infty} \mathscr{P}(\tau_k(x, y) \geq i | X_k = x, X'_k = y) < \infty.$$

□

In the literature, the well-known Hajnal weak ergodicity coefficient is defined by (in our notation) (Park et al. 1993; Hajnal 1958)

$$\kappa := \sup_k \Big(1 - \inf_{x,x' \in \mathscr{S}_k} \{ \sum_{y \in \mathscr{S}_{k,out}} \min[P_k(y|x), P_k(y|x')] \} \Big).$$

When $\mathscr{S}_{k,out}$ is finite, this is

$$\kappa = 1 - \inf_k \Big(\min_{x,x'} \{ \sum_y \min[P_k(y|x), P_k(y|x')] \} \Big).$$

Lemma 2.3 $\kappa < 1$ *if and only if* $\nu < 1$.

Proof The "If" part $\Longleftarrow$: Because $\sum_y \min[P_k(y|x), P_k(y|x')] \geq \sum_y [P_k(y|x) P_k(y|x')]$, we have $\kappa \leq \nu$. Thus, $\kappa < 1$ if $\nu < 1$.

The "Only if" part $\Longrightarrow$: We prove if $\nu = 1$ then $\kappa = 1$. Equivalently, if

$$\inf_k \Big(\min_{x,x' \in \mathscr{S}_k} \{ \sum_{y \in \mathscr{S}_{k,out}} [P_k(y|x) P_k(y|x')] \} \Big) = 0, \tag{2.6}$$

then

$$\inf_k \Big(\min_{x,x'} \{ \sum_y \min[P_k(y|x), P_k(y|x')] \} \Big) = 0. \tag{2.7}$$

By (2.6), there is a sequence of integers $k_1, k_2, \ldots$ (not necessary in increasing order), and a sequence of positive numbers ε_l, with $\lim_{l\to\infty} \varepsilon_l = 0$, such that

$$\min_{x,x' \in \mathscr{S}_{k_l}} \{ \sum_{y \in \mathscr{S}_{k_l,out}} [P_{k_l}(y|x) P_{k_l}(y|x')] \} < \varepsilon_l.$$

Thus, there is a pair denoted by x_l and x_l' such that

$$\sum_{y \in \mathscr{S}_{k_l,out}} [P_{k_l}(y|x_l) P_{k_l}(y|x_l')] < \varepsilon_l.$$

So, $P_{k_l}(y|x_l) P_{k_l}(y|x_l') < \varepsilon_l$ for all $y \in \mathscr{S}_{k_l,out}$. Furthermore,

$$\{\min[P_{k_l}(y|x_l), P_{k_l}(y|x_l')]\}^2 \le P_{k_l}(y|x_l) P_{k_l}(y|x_l') < \varepsilon_l,$$

or

$$\min[P_{k_l}(y|x_l), P_{k_l}(y|x_l')] < \sqrt{\varepsilon_l},$$

for all $y \in \mathscr{S}_{k_l,out}$. Therefore,

$$\sum_y \min[P_k(y|x_l), P_k(y|x_l')] < M_1 \sqrt{\varepsilon_l},$$

where M_1 is the bound of the sizes of $\mathscr{S}_k$, $k = 0, 1, \ldots$; and thus,

$$\min_{x,x'} \{\sum_y \min[P_k(y|x), P_k(y|x')]\} < M_1 \sqrt{\varepsilon_l}.$$

Finally, there is a sequence, $k_1, k_2, \ldots$, such that

$$\min_{x,x'} \{\sum_y \min[P_k(y|x_l), P_k(y|x_l')]\} < M_1 \sqrt{\varepsilon_l} \to 0.$$

This implies (2.7). □

Therefore, Hajnal weak ergodicity with $\kappa < 1$ is equivalent to confluencity with $\nu < 1$, which implies strong confluencity. By Lemma 2.1, confluencity implies weak ergodicity, but the next example shows that the opposite is not true; Fig. 2.1 shows the relations.

Example 2.1 The state space $\mathscr{S}_k$ consists of two states denoted by $\mathscr{S}_k := \{1_k, 2_k\}$, $k = 0, 1, \ldots$. Let $P(1_1|1_0) = P(2_1|1_0) = P(1_1|2_0) = P(2_1|2_0) = 0.5$, $P(1_{k+1}|1_k) = 1$, $k = 1, 2, \ldots$, and $P(2_{k+1}|2_k) = 1$, $k = 1, 2, \ldots$; see Fig. 2.2. It is clear that states 1_0 and 2_0 are not confluent to each other (two independent sample paths

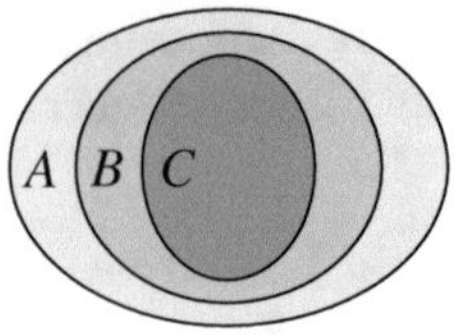

Fig. 2.1 Confluencity versus weak ergodicity. A: Weak ergodicity; B: Confluencity; C: Confluencity with $\nu < 1$ and weak ergodicity with $\kappa < 1$. A–C: $\nu = 1$ and $\kappa = 1$

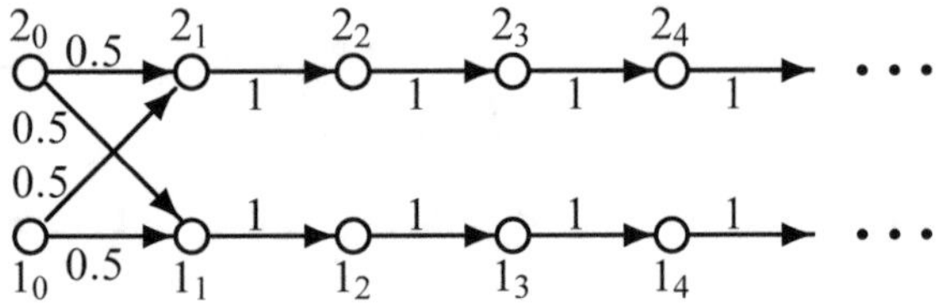

Fig. 2.2 The state transitions in Example 2.1

starting from 1_0 and 2_0 have a probability of only 0.5 to meet), i.e., (2.2) does not hold. However, the weak ergodicity condition (2.4) does hold for $x = 1_0$ and $x' = 2_0$.

As expected, in this example, we have $\nu = \kappa = 1$. □

2.2 State Classification and Decomposition

State classification is important in characterizing the relations among the states, and the optimality conditions for different classes are usually different (cf. Puterman 1994; Cao 2007 for THMCs). However, notions such as connectivity, aperiodicity, recurrent and transient states, which play important roles in state classification of THMCs, no longer apply to TNHMCs. In this section, we show that with confluencity, state classification can be carried out for TNHMCs and all the states in a TNHMC can be grouped into a number of confluent and branching classes. This extends the recurrent-transient state decomposition of THMCs.

Naturally, the confluencity-based classification also applies to THMCs. Confluent and branching are similar to, but different from, recurrent and transient in THMCs (cf. Example 2.5). We will see that, compared with recurrent and transient states, the notion of confluent and branching states is more appropriate for state classification and performance optimization.

2.2.1 Connectivity of States

Consider two independent sample paths starting from k, $\{X_l, l \geq k\}$ and $\{X'_l, l \geq k\}$, with the same transition law $\mathbb{P} = \{P_l, l \geq k\}$ and state spaces $\mathbb{S} = \{\mathscr{S}_l, l \geq k\}$, but with two different initial states $X_k = x$, $X'_k = y$, $x, y \in \mathscr{S}_k$.

Definition 2.3 (*Connectivity of states*) If Eq. (2.2) holds, i.e.,

$$\lim_{K\to\infty} \mathscr{P}(\tau_k(x,y) > K | X_k = x, X'_k = y) = 0,$$

for two states $x, y \in \mathscr{S}_k$ at k, $k = 0, 1, \ldots$, then these two states x and y at time k are said to be *connected*, and this connectivity is denoted by $x \rtimes y$; Two states x and y at time k are said to be *strongly connected*, if in addition, we have $E[\tau_k(x,y)|X_k = x, X'_k = y] < \infty$.[1] □

From the definition, connectivity is not a new notion, and it is introduced simply for convenience; two connected states are confluent to each other.

Lemma 2.4 *Connectivity is reflective, symmetric, and transitive; i.e.,*

(a) $x \rtimes x$,
(b) if $x \rtimes y$ *then* $y \rtimes x$, *and*
(c) if $x \rtimes y$ *and* $y \rtimes z$, *then* $x \rtimes z$, *for* $x, y, z \in \mathscr{S}_k$, $k = 0, 1, \ldots$.

Proof Items (a) and (b) follow directly from the definition. To prove c), we consider three independent sample paths, $\{X_l^x, l \geq k\}$, $\{X_l^y, l \geq k\}$, and $\{X_l^z, l \geq k\}$, starting from states $X_k^x = x$, $X_k^y = y$, and $X_k^z = z$, respectively. First, we note that $\tau_k(x, y)$ in (2.1) is a random variable representing the confluent time of $\{X_l^x, l \geq k\}$ and $\{X_l^y, l \geq k\}$, and $\tau_k(y,z)$ is the confluent time of $\{X_l^y, l \geq k\}$ and $\{X_l^z, l \geq k\}$. By definition, we have $X^x_{\tau_k(x,y)} = X^y_{\tau_k(x,y)}$. Because of the Markov property, the two random sequences $\{X^x_{\tau_k(x,y)}, X^x_{\tau_k(x,y)+1}, \ldots\}$ and $\{X^y_{\tau_k(x,y)}, X^y_{\tau_k(x,y)+1}, \ldots\}$ are identically distributed. Therefore, conditioned on $\tau_k(y,z) \geq \tau_k(x,y)$ and $\tau_k(x,z) \geq \tau_k(x,y)$, the random variables $\tau_k(x,z)$ and $\tau_k(y,z)$ have the same distribution; thus,

$$\begin{aligned}
&\mathscr{P}(\tau_k(x,z) > K | \tau_k(y,z) \geq \tau_k(x,y)) \\
&= \mathscr{P}[\tau_k(x,z) > K | \tau_k(y,z) \geq \tau_k(x,y), \tau_k(x,z) \geq \tau_k(x,y)] \\
&\quad + \mathscr{P}[\tau_k(x,z) > K | \tau_k(y,z) \geq \tau_k(x,y), \tau_k(x,z) < \tau_k(x,y)] \\
&= \mathscr{P}[\tau_k(y,z) > K | \tau_k(y,z) \geq \tau_k(x,y), \tau_k(x,z) \geq \tau_k(x,y)] \\
&\quad + \mathscr{P}[\tau_k(x,z) > K | \tau_k(y,z) \geq \tau_k(x,y), \tau_k(x,z) < \tau_k(x,y)]. \qquad (2.8)
\end{aligned}$$

Because $\lim_{K\to\infty} \mathscr{P}(\tau_k(y,z) > K) = 0$ and the probability that both $\tau_k(y,z) \geq \tau_k(x,y)$ and $\tau_k(x,z) < \tau_k(x,y)$ is positive, we have

$$\lim_{K\to\infty} \mathscr{P}[\tau_k(y,z) > K | \tau_k(y,z) \geq \tau_k(x,y), \tau_k(x,z) < \tau_k(x,y)] = 0.$$

Next, given $\tau_k(x,z) < \tau_k(x,y)$, the fact $\tau_k(x,z) > K$ implies $\tau_k(x,y) > K$; thus,

[1] We also say that "these two states are confluent to each other," but this statement may create a bit of confusion. Precisely, in this sense, "the two states are confluent (to each other)" is different from "the two states are confluent states" (see Example 2.3). Therefore, connectivity is not a new concept, and it is introduced simply for clarity and convenience.

$$\mathscr{P}(\tau_k(x,z) > K | \tau_k(y,z) \geq \tau_k(x,y), \tau_k(x,z) < \tau_k(x,y))$$
$$\leq \mathscr{P}[\tau_k(x,y) > K | \tau_k(y,z) \geq \tau_k(x,y), \tau_k(x,z) < \tau_k(x,y)].$$

By (2.2), we have

$$\lim_{K\to\infty} \mathscr{P}(\tau_k(x,z) > K | \tau_k(y,z) \geq \tau_k(x,y), \tau_k(x,z) < \tau_k(x,y)) = 0.$$

Therefore, from (2.8), we get

$$\lim_{K\to\infty} \mathscr{P}(\tau_k(x,z) > K | \tau_k(y,z) \geq \tau_k(x,y)) = 0. \tag{2.9}$$

When $\tau_k(y,z) < \tau_k(x,y)$, we exchange the role of x and z in the above analysis, and obtain

$$\lim_{K\to\infty} \mathscr{P}(\tau_k(x,z) > K | \tau_k(y,z) < \tau_k(x,y)) = 0.$$

From this equation and (2.9), we conclude that $\lim_{K\to\infty} \mathscr{P}(\tau_k(x,z) > K) = 0$, i.e., $x \rightarrowtail\!\!\!\!\leftarrowtail z$. □

Intuitively, we may say that two independent sample paths starting from two connected states meet in a finite time w.p.1. We also say that two states at two different times are connected if two independent sample paths starting from these two states meet in a finite time w.p.1.

2.2.2 *Confluent States*

For any two connected states x and y, there is an issue whether the two sample paths starting from each of them, respectively, will meet infinitely often, or will never meet again after meeting a few times. Exploring this property leads to state classification. First, we say a state $y \in \mathscr{S}_{k'}$ is *reachable* from $x \in \mathscr{S}_k$, $k' > k$, if $\mathscr{P}(X_{k'} = y | X_k = x) > 0$.

Definition 2.4 A state $x \in \mathscr{S}_k$ is said to be a confluent state if any two states at any time $k' > k$ that are reachable from $x \in \mathscr{S}_k$ are connected; otherwise, it is said to be a branching state. □

Intuitively speaking, two independent sample paths starting from a confluent state meet infinitely often w.p.1; and two independent sample paths starting from two connected states meet at least once. By convention, if, starting from a state, eventually the chain can reach only one state at every time, this state is confluent and is called an *absorbing state*, see Fig. 2.6 (cf. the same notion for THMCs Puterman 1994).

Example 2.2 States 1_0 (or 2_0) in Fig. 2.2 is not a confluent state because its downstream states 1_1 and 2_1 are not connected, or confluent, to each other. On the other

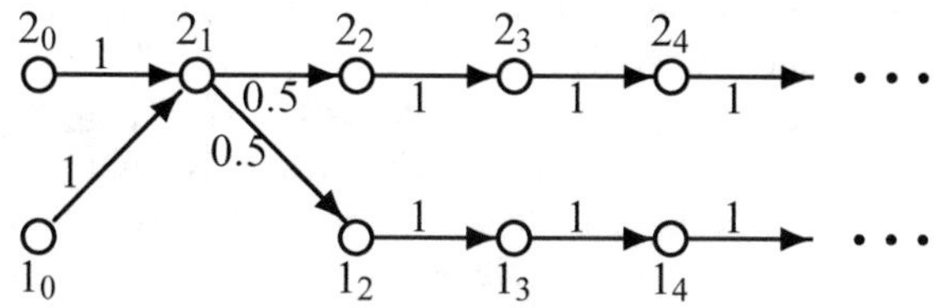

Fig. 2.3 The state transitions in Example 2.3

hand, both states 1_1 and 2_1 are absorbing states, and by definition, they are confluent states. States 1_0 and 2_0 are branching states, and each of them is "branching" into the two absorbing states. □

Example 2.3 In Fig. 2.3, two states 1_0 and 2_0 are connected, or confluent, to each other; but both of them are not confluent states. The two sample paths starting from 1_0, or 2_0, meet once at time $k = 1$, and they may never meet again afterward. States $1_2, 1_3, \ldots$ and $2_2, 2_3, \ldots$ are two confluent classes (absorbing states), and states 1_0, 2_0, and 2_1 are branching. □

In summary, all the states can be classified into two types of states, confluent and branching. Furthermore, from Lemma 2.4, we may group the confluent states in $\mathscr{S}_k$ that are connected to each other together. Each group is called a *confluent class* at time k; each confluent class consists of all the confluent states that are connected to each other. In addition, if two confluent states at different times are connected, we also say that they belong to the same confluent class. Denote the confluent classes at time k by $\mathscr{R}_{k,1}$, $\mathscr{R}_{k,2}$, ..., $\mathscr{R}_{k,d_k}$, $k = 0, 1, \ldots$, with d_k being the number of the confluent classes at k.

Lemma 2.5 *(a) Any sample path starting from a confluent state passes through only confluent states;*

(b) If two states $X_k = x$ and $X'_k = y$ are confluent states and connected to each other, then at any time $k' > k$, the states $X_{k'} = u$ and $X'_{k'} = v$ are also confluent states and connected to each other;

(c) Two independent sample paths starting from any two states in different confluent classes cannot meet with any positive probability.

Proof (a) Let $\{X_l^x, l \ge k\}$ be a sample path with $X_k^x = x$ being a confluent state. Suppose that at a time $k' > k$, $X_{k'}^x = y$, and y is not a confluent state. Consider two independent sample paths $\{X_l^y, l \ge k'\}$ and $\{X'^y_l, l \ge k'\}$ with $X_{k'}^y = X'^y_{k'} = y$. Because y is branching, and by Definition 2.4, there is a $k'' > k'$ such that with a positive probability the two states $X_{k''}^y = u$ and $X'^y_{k''} = v$ are not connected. Note that every sample path $\{X_l^y, l \ge k'\}$ starting from $k' > k$ is a part of a sample path $\{X_l^x, l \ge k\}$ starting from k; furthermore, the probability that the sample path $\{X_l, l \ge k\}$ reaches $X_{k'} = y$ is positive. Therefore, with a positive probability, the two states $X_{k''}^x = u$ and $X'^x_{k''} = v$ are not connected. This contradicts the fact that x at k is a confluent state.

(b) Following (a), both states $X_{k'} = u$ and $X'_{k'} = v$ are confluent states. Let two independent sample paths $\{X_l^x, l \geq k\}$ and $\{X'^x_l, l \geq k\}$ meet at $k'' > k$ with $X^x_{k''} = X'^x_{k''} = z$. If $k'' \geq k'$, then $\{X_l^u, l \geq k'\}$ and $\{X'^v_l, l \geq k'\}$ meet in a finite time at k''. On the other hand, if $k'' < k'$, then because z is confluent, $\{X_l^u, l \geq k'\}$ and $\{X'^v_l, l \geq k'\}$ must also meet in a finite time after k'. Thus, u and v are connected.

(c) Suppose the opposite is true. That is, two independent sample paths, $\{X_l, l \geq k\}$ with $X_k = x \in \mathscr{R}_{k,1}$ and $\{X'_l, l \geq k\}$ with $X'_k = y \in \mathscr{R}_{k,2}$, meet with a positive probability $\varepsilon > 0$ at state u at some time $k' > k$, $X_{k'} = X'_{k'} = u$. Pick up any state $z \in \mathscr{R}_{k,1}$ and suppose another independent sample path $\{X''_l, l \geq k\}$ with $X''_k = z$ reaches state v at k', $X''_{k'} = v$. By part b), $X''_{k'} = v$ and $X_{k'} = u$ are connected. Because z is any state in $\mathscr{R}_{k,1}$, this means that every state in $\mathscr{S}_{k'}$, which is reachable from any state in $\mathscr{R}_{k,1}$, and state u at k' are connected. Similarly, every state in $\mathscr{S}_{k'}$, which is reachable from any state in $\mathscr{R}_{k,2}$, and state u at k' are connected. By Lemma 2.4, every state in $\mathscr{S}_{k'}$ reachable from any state in $\mathscr{R}_{k,1}$ and every state in $\mathscr{S}_{k'}$ reachable from any state in $\mathscr{R}_{k,2}$ are connected. Therefore, every state in $\mathscr{R}_{k,1}$ and every state in $\mathscr{R}_{k,2}$ must be connected. That is, $\mathscr{R}_{k,1}$ and $\mathscr{R}_{k,2}$ must be in the same confluent class. This is not correct. Therefore, (c) must be true. □

From Lemma 2.5(b) and (c), independent sample paths starting from different confluent states either meet w.p.1 or with probability zero. The sample paths starting from states in different confluent classes do not interact at all. New states may join the existing confluent classes, or form new confluent classes. Thus, the number of confluent classes may increase and will not decrease. However, under different policies, the numbers of confluent classes may differ, even if the numbers of confluent states are the same. Let

$$d \leq \max_k S_k < \infty$$

be the number of all the confluent classes for states at all $k = 0, 1, \ldots$. Then, d_k increases to d, as $k \to \infty$. Denote the confluent classes by $\mathscr{R}_{\cdot r}$, $r = 1, 2, \ldots, d$; then,

$$\mathscr{R}_{k,r} := \mathscr{R}_{\cdot r} \cap \mathscr{S}_k \text{ and } \mathscr{R}_k := \cup_{r=1}^{d_k} \mathscr{R}_{k,r} \tag{2.10}$$

are the set of all confluent states in the rth class at k and the set of all confluent states at k, respectively, and

$$\mathscr{R} := \cup_{k=0}^{\infty} \mathscr{R}_k \tag{2.11}$$

is the set of all confluent states.

Figure 2.4 illustrates two confluent classes $\mathscr{R}_1$ and $\mathscr{R}_2$, with $x_0, x_1, x_2, x_3, x_4 \in \mathscr{R}_1$, and $x_5, x_6, x_7 \in \mathscr{R}_2$.

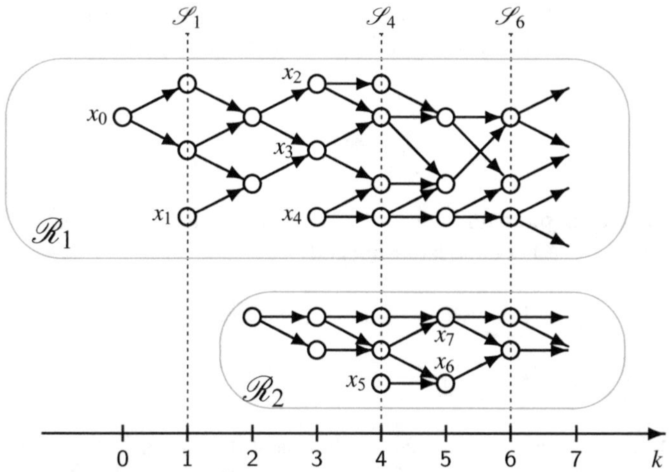

Fig. 2.4 Two confluent classes, $\mathscr{R}_1$ and $\mathscr{R}_2$

2.2.3 Branching States

According to Definitions 2.3 and 2.4, for a branching state, there exists a finite time $K > 0$ such that after time K any two independent sample paths starting from this state will meet with only a positive probability of less than one (<1).

The set of branching states at time k is denoted by $\mathscr{T}_k$, and the set of branching states at all times is then

$$\mathscr{T} = \cup_{k=0}^{\infty} \mathscr{T}_k. \tag{2.12}$$

In a finite state TNHMC, a sample path from a branching state eventually reaches one of at least two confluent (including absorbing) classes. Let

$$\tau_{k,\mathscr{R}}(x) := \min\{\tau \geq k, X_\tau \in \mathscr{R}\} - k, \qquad x \in \mathscr{T}_k.$$

Then, we have

$$\lim_{K\to\infty} \mathscr{P}(\tau_{k,\mathscr{R}}(x) > K | X_k = x) = 0. \tag{2.13}$$

Sometimes, however, we may need a branching state to be *strongly branching*, defined by

$$E[\tau_{k,\mathscr{R}}(x) | X_k = x] \\ = \sum_{K=1}^{\infty} K\{\mathscr{P}(\tau_{k,\mathscr{R}}(x) = K | X_k = x)\} < \infty. \tag{2.14}$$

Let $p_{k,r}(x), x \in \mathscr{T}_k$ and $r = 1, 2, \ldots, d$, be the probability that a sample path starting from x at time k will eventually join the rth confluent class, with

$$\sum_{r=1}^{d} p_{k,r}(x) = 1.$$

Because once a sample path enters a confluent class, it will stay there forever, so $\mathscr{P}(X_K \in \mathscr{R}_r | X_k = x)$ is nondecreasing and

$$\lim_{K \to \infty} \mathscr{P}(X_K \in \mathscr{R}_r | X_k = x) = p_{k,r}(x), \tag{2.15}$$

for all $x \in \mathscr{T}_k, r = 1, 2, \ldots, d$.

Definition 2.5 A Markov chain $\boldsymbol{X}$, or its transition law $\mathbb{P}$, is said to be strongly connected, if $E[\tau_k(x, y)|X_k = x, X'_k = y] < M_3 < \infty$ for all confluent states in the same class $x, y \in \mathscr{R}_{k,r}, r = 1, 2, \ldots, d_k, k = 1, 2, \ldots$, and $E[\tau_{x,\mathscr{R}}|X_k = x] < M_3 < \infty$ for all branching states $x \in \mathscr{T}_k, k = 0, 1, 2, \ldots$.

If the number of states $S_k, k = 0, 1, \ldots$, is not bounded, a sample path may branch out endlessly and never join in any confluent class. We will not discuss this case since S_k is bounded in this book.

Figure 2.5 illustrates a typical situation of the branching states. Sample paths starting from states x_0 will either join the confluent class $\mathscr{R}_1$ at states x_5 or x_6, or join another confluent class $\mathscr{R}_2$ at x_7 or x_8. All the (red) states in set $\mathscr{T}$ are branching.

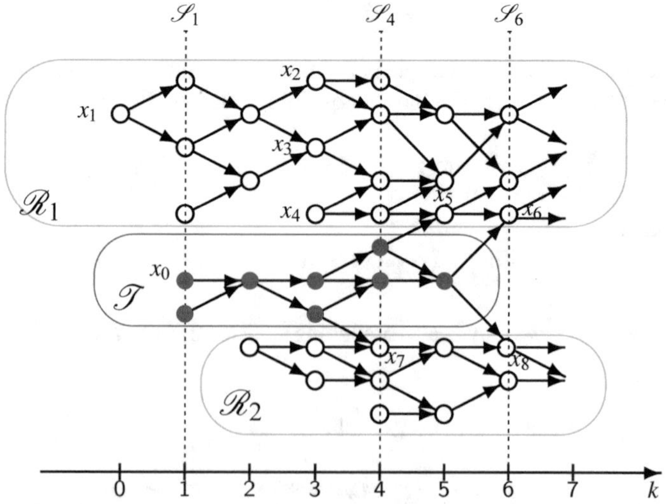

Fig. 2.5 The branching states in set $\mathscr{T}$ and two classes of confluent states in sets $\mathscr{R}_1$ and $\mathscr{R}_2$

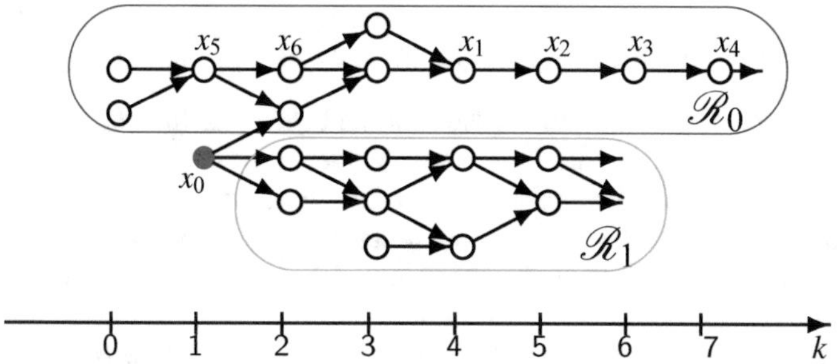

Fig. 2.6 Two confluent classes $\mathscr{R}_0$ and $\mathscr{R}_1$, all states in $\mathscr{R}_0$ are absorbing, x_0 is branching

In Fig. 2.6, all the states in $\mathscr{R}_0$, including states x_1 to x_6, are absorbing; in particular, starting from x_5 or x_6, the chain eventually reaches only one state, x_1, x_2, x_3, or x_4, at any time. These states form a special confluent class. There is another confluent class $\mathscr{R}_1$. State x_0 is branching since it may join either one of these two confluent classes.

2.2.4 State Classification and Decomposition Theorem

We may summarize the above results in the following theorem.

Theorem 2.1 (State Classification and Decomposition Theorem) *Let the state space at time k of a TNHMC be $\mathscr{S}_k$, $k = 0, 1, \ldots$, and $\mathscr{S} := \cup_{k=0}^{\infty} \mathscr{S}_k$. Assume $S_k = |\mathscr{S}_k| < M_1 < \infty$, for all k. The following state decomposition properties hold:*

1. *All the states belong to one of two types, the confluent states $\mathscr{R}$ or the branching states $\mathscr{T}$, with $\mathscr{S} = \mathscr{R} \cup \mathscr{T}$, $\mathscr{R} = \cup_{k=0}^{\infty} \mathscr{R}_k$, $\mathscr{T} = \cup_{k=0}^{\infty} \mathscr{T}_k$, $\mathscr{R}_k = \mathscr{S}_k \cap \mathscr{R}$, $\mathscr{T}_k = \mathscr{S}_k \cap \mathscr{T}$, and $\mathscr{S}_k = \mathscr{R}_k \cup \mathscr{T}_k$.*
2. *All the confluent states at time k can be grouped into d_k confluent classes $\mathscr{R}_{k,r}$, $r = 1, 2, \ldots, d_k$, with $\mathscr{R}_k = \cup_{r=1}^{d_k} \mathscr{R}_{k,r}$ and $\mathscr{R}_{k,r} \cap \mathscr{R}_{k,r'} = \emptyset$, $r \neq r'$.*
   a. *Two independent sample paths starting from the same confluent state, or from any two states in the same confluent class, meet infinitely often w.p.1, and two independent sample paths starting from any two states in two different confluent classes will never meet;*
   b. *All sample paths starting from the states in the same confluent class at time k, $k = 0, 1, \ldots$, pass through the states in the same confluent class at all $k' > k$;*
   c. *The number of confluent classes at time k, d_k, is nondecreasing with maximum number $d \leq \max_k S_k$, i.e., if k is large enough, then $d_k = d$; and*
3. *A sample path starting from any branching state $x \in \mathscr{T}_k$, $k = 0, 1, \ldots$, will eventually reach one of at least two confluent classes with probability one; denoting*

the probability that a sample path starting from $x \in \mathscr{T}_k$ will reach the rth confluent class by $p_{k,r}(x)$, $r = 1, 2, \ldots, d$, we have $\sum_{r=1}^{d} p_{k,r}(x) = 1$.

By Theorem 2.1, we may number the confluent classes at time k, $\mathscr{R}_{k,1}, \ldots, \mathscr{R}_{k,d_k}$, in such a way that the states in $\mathscr{R}_{k,r}$ can only reach the states in $\mathscr{R}_{k',r}$, for the same r, $r = 1, 2, \ldots, d_k$, and for any $k' > k$. In this way, all the sample paths starting from confluent states can be grouped into d classes; any sample path in the rth class, $r = 1, 2, \ldots, d$, travels through the states in the rth confluent states, i.e., remains in $\mathscr{R}_{k,r}$, $k = 0, 1, \ldots$.

A TNHMC is called a *uni-chain* if $d = 1$, or a *multi-class* if $d > 1$ (cf. Cao 2007 and Puterman 1994 for the case of THMCs). The performance optimization of uni-chain and multi-class TNHMCs is discussed in Chaps. 3 and 4, respectively. Applying Lemma 2.1 to every confluent class in a multi-class Markov chain, we have

Theorem 2.2 *For a multi-class TNHMC with $S_k < M_1 < \infty$, $k = 1, 2, \ldots$, the weak ergodicity holds for every class of confluent states; i.e., for any $x \in \mathscr{R}_{k,r}$ and $x' \in \mathscr{R}_{k',r}$, $r = 1, 2, \ldots, d_k$, and any $k, k' = 0, 1, \ldots$, we have*

$$\lim_{K \to \infty} \left\{ \max_{y \in \mathscr{S}_K} \left| \mathscr{P}(X_K = y | X_k = x) - \mathscr{P}(X_K = y | X_{k'} = x') \right| \right\} = 0,$$

in which $\mathscr{P}(X_K = y | X_k = x) > 0$ only if y is in the same confluent class as x, i.e., $y \in \mathscr{R}_{K,r}$.

As shown in the next two examples, the notions of confluent and branching are similar to, but different from, that of recurrent and transient in THMCs. Specifically, recurrent states in THMCs are confluent, but transient states may or may not be branching. If from a transient state, the THMC finally joins only one recurrent class, the transient state is confluent, not branching. Thus, in a uni-chain THMCs, all the states are confluent, even some of them may be transient. It is well known that the performance optimization of a uni-chain THMC consisting one recurrent class and transient states is the same as that of an ergodic chain, which consists of only recurrent states (Cao 2007; Puterman 1994). Thus, compared with recurrent and transient, the notion of confluent and branching is more appropriate for state classification and performance optimization.

Example 2.4 In a THMC with the transition probability matrix

$$P = \begin{bmatrix} 0.5 & 0.5 & 0 \\ 0.5 & 0.5 & 0 \\ 0.5 & 0 & 0.5 \end{bmatrix},$$

all the sample paths starting with any state 1, 2, or 3 eventually stay in the set $\{1, 2\}$, so states 1 and 2 are recurrent, and state 3 is transient. However, all these three states are confluent. □

Table 2.1 The state classification of TNHMCs and THMCs

	State space	
TNHMCs	Branching states	Confluent states
THMCs	Transient states	Recurrent states

Example 2.5 Consider a THMC with the transition probability matrix

$$P = \begin{bmatrix} 0.5 & 0.5 & 0 & 0 & 0 & 0 & 0 \\ 0.5 & 0.5 & 0 & 0 & 0 & 0 & 0 \\ 0.5 & 0 & 0.5 & 0 & 0 & 0 & 0 \\ 0 & 0 & 0 & 0.5 & 0.5 & 0 & 0 \\ 0 & 0 & 0 & 0.5 & 0.5 & 0 & 0 \\ 0 & 0 & 0 & 0.5 & 0 & 0.5 & 0 \\ 0 & 0 & 0.25 & 0 & 0 & 0.25 & 0.5 \end{bmatrix}.$$

There are two recurrent classes, consisting of states 1 and 2, and states 4 and 5, respectively. So states 1, 2, 4, and 5 are recurrent. There are three transient states, 3, 6, and 7. However, in the terminology of TNHMCs, there are two classes of confluent states, 1, 2, 3, and 4, 5, 6; in particular, states 3 and 6 are confluent states, and they are no different from any other confluent state; only state 7 is branching since starting from it the Markov chain may join two different confluent classes. □

Table 2.1 illustrates the comparison of the state classification of TNHMCs and that of THMCs; it shows that a transient state may be either a branching state or a confluent state, and a confluent state may be either transient (e.g., in a uni-chain) or recurrent.

In the literature, there is a decomposition–separation (DS) theorem for sample paths of TNHMCs (Blackwell 1945; Cohn 1989; Sonin 1991, 1996, 2008). It shows that any sample path of a TNHMC will, after a finite number of steps, enter one of the jets (a jet is a sequence of subsets of the state spaces at time $k = 0, 1, \ldots$), and weak ergodicity holds for every jet. This theorem can be easily derived and clearly explained by Theorems 2.1 and 2.2; see Cao (2016) for more discussion.

Chapter 3
Optimization of Average Rewards and Bias: Single Class

In this chapter, we study the optimization of the long-run average and bias of single class (or uni-chain) TNHMCs.

In Sect. 3.2, with confluencity, we define the relative performance potentials, from which we define the most central notion in performance optimization, the performance potentials, and discuss its properties. In Sect. 3.3, with the performance potentials, we derive the difference formula for the average rewards of any two policies; and based on which we obtain the necessary and/or sufficient optimality conditions for average rewards. In Sect. 3.4, we study the bias optimality, which optimizes the transient performance in the initial period. Bias potentials are defined, and bias optimality conditions are derived.

The analysis is based on the relative optimization approach, which is based on a performance difference formula that gives the difference of the performance measures of any two policies on the entire infinite horizon. The under-selectivity is reflected in the optimality conditions, for both the average reward and bias (Theorems 3.2 and 3.5); it is clear that the optimality conditions do not need to hold in any finite period, or in any "non-frequently" visited sequence of time instants.

3.1 Preliminary

The long-run average is defined by the "lim inf" in (1.6)

$$\eta_k(x) = \liminf_{K\to\infty} \frac{1}{K} E\Big\{ \sum_{l=k}^{k+K-1} f_l(X_l) \Big| X_k = x \Big\}. \tag{3.1}$$

If the limit exists, then the long-run average becomes (1.7):

X.-R. Cao, *Foundations of Average-Cost Nonhomogeneous Controlled Markov Chains*, SpringerBriefs in Control, Automation and Robotics,
https://doi.org/10.1007/978-3-030-56678-4_3

$$\eta_k(x) = \lim_{K\to\infty} \frac{1}{K} E\Big\{ \sum_{l=k}^{k+K-1} f_l(X_l) \Big| X_k = x \Big\}.$$

In this case, because the operator "lim" is linear and additive, we have

$$\eta_k = P_k \eta_{k+1}. \tag{3.2}$$

First, we define some technical terms. Let a_k, $k = 0, 1, \ldots$, be a sequence of real numbers.

Definition 3.1 A subsequence a_{k_l}, $l = 0, 1, \ldots$, is called an *inf-subsequence* of a_k, $k = 0, 1, \ldots$, if

$$\liminf_{l\to\infty} a_{k_l} = \liminf_{k\to\infty} a_k.$$

It is called a *lim-subsequence* of a_k, $k = 0, 1, \ldots$, if $\lim_{l\to\infty} a_{k_l}$ exists and

$$\lim_{l\to\infty} a_{k_l} = \liminf_{k\to\infty} a_k. \qquad \square$$

Consider two sequences of real numbers a_k and b_k. Let a_{k_l} and $b_{k'_l}$, $l = 0, 1, \ldots$, be their lim-subsequences, respectively. These two sequences k_l and k'_l, $l = 0, 1, \ldots$, may be completely different; and therefore, in general, we have

$$\liminf_{k\to\infty} (a_k + b_k) \geq \liminf_{k\to\infty} a_k + \liminf_{k\to\infty} b_k. \tag{3.3}$$

Because of (3.3), in general, (3.2) may not hold for η_k defined in (3.1) with "lim inf".

However, if the confluencity holds, the sample paths starting from different states will eventually totally mix together. We can prove that for uni-chain TNHMCs, $\eta_k(x)$ defined in (3.1) by "lim inf" does not depend on k and x, and therefore, (3.2) holds. First, we have

Lemma 3.1 *For any two sequences of real numbers a_k and b_k, $k = 0, 1, \ldots$, if $\lim_{k\to\infty} a_k$ exists, then*

$$\liminf_{k\to\infty} (a_k + b_k) = \lim_{k\to\infty} a_k + \liminf_{k\to\infty} b_k.$$

Proof It can be verified by the definition of "lim inf" directly. $\square$

Lemma 3.2 *In a uni-chain TNHMC, under Assumption 1.1, the long-run averages defined in (3.1) (with* "lim inf"*) are the same for all the states at all times; i.e.,*

$$\eta_k(x) \equiv \eta, \ \forall\, x \in \mathscr{S}_k, \ k = 0, 1, \ldots. \tag{3.4}$$

Proof Consider two independent sample paths of the same uni-chain TNHMC, denoted by $\{X_l, l = 0, 1, \ldots\}$ and $\{X'_l, l = 0, 1, \ldots\}$, starting from two different states at k, $X_k = x$ and $X'_k = y$. With the confluencity (2.2), for any $\varepsilon > 0$, there is a large enough K_0, such that

$$\mathscr{P}(\tau_k(x, y) > K_0 | X_k = x, X'_k = y) < \varepsilon. \tag{3.5}$$

Then, we have

$$\begin{aligned}
&\sum_{l=k}^{k+K-1} E\{[f(X'_l) - f(X_l)] | X'_k = y, X_k = x\} \\
= &\sum_{l=k}^{k+K-1} \Big\{ E\{[f(X'_l) - f(X_l)] | X'_k = y, X_k = x, \tau_k(x, y) \leq K_0\} \\
&\qquad \mathscr{P}(\tau_k(x, y) \leq K_0 | X_k = x, X'_k = y) \\
&\quad + E\{[f(X'_l) - f(X_l)] | X'_k = y, X_k = x, \tau_k(x, y) > K_0\} \\
&\qquad \mathscr{P}(\tau_k(x, y) > K_0 | X_k = x, X'_k = y) \Big\}.
\end{aligned}$$

By the Markov property, we have $E\{[f(X'_l) - f(X_l)] | X'_k = y, X_k = x, \tau_k(x, y) \leq K_0\} = 0$, for $l \geq k + K_0$. Therefore, in the first term on the right-hand side of the above equation, we have

$$\begin{aligned}
&\sum_{l=k}^{k+K-1} \Big\{ E\{[f(X'_l) - f(X_l)] | X'_k = y, X_k = x, \tau_k(x, y) \leq K_0\} \\
= &\sum_{l=k}^{k+K_0-1} \Big\{ E\{[f(X'_l) - f(X_l)] | X'_k = y, X_k = x, \tau_k(x, y) \leq K_0\} \\
\leq &2K_0 M_2.
\end{aligned}$$

By Assumption 1.1(b) and (3.5), we have

$$\begin{aligned}
&\Big| \sum_{l=k}^{k+K-1} E\{[f(X'_l) - f(X_l)] | X'_k = y, X_k = x, \tau_k(x, y) > K_0\} \\
&\qquad \mathscr{P}(\tau_k(x, y) > K_0 | X_k = x, X'_k = y) \Big| < 2K M_2 \varepsilon.
\end{aligned}$$

From the above three equations, we have

$$\lim_{K \to \infty} \frac{1}{K} \Big| \sum_{l=k}^{k+K-1} E\{[f(X'_l) - f(X_l)] | X'_k = y, X_k = x\} \Big| < 2M_2 \varepsilon.$$

Because ε can be any positive number so long as $K_0 (< K)$ is large enough, we have

$$\lim_{K\to\infty}\frac{1}{K}\left|\sum_{l=k}^{k+K-1}E\{[f(X'_l)-f(X_l)]|X'_k=y,X_k=x\}\right|=0.$$

Next, we have

$$\begin{aligned}&\frac{1}{K}\sum_{l=k}^{k+K-1}E\{f(X'_l)|X'_k=y\}\\ =&\frac{1}{K}\sum_{l=k}^{k+K-1}E\{[f(X'_l)-f(X_l)]|X'_k=y,X_k=x\}\\ &+\frac{1}{K}\sum_{l=k}^{k+K-1}E\{f(X_l)|X_k=x\}.\end{aligned}$$

Now, we set

$$a_K:=\frac{1}{K}\sum_{l=k}^{k+K-1}E\{[f(X'_l)-f(X_l)]|X'_k=y,X_k=x\}$$

and

$$b_K:=\frac{1}{K}\sum_{l=k}^{k+K-1}E\{f(X_l)|X_k=x\}.$$

By Lemma 3.1, we can prove $\eta_k(x)=\eta_k(y)$; i.e., $\eta_k(x)$ does not depend on $x\in\mathscr{S}_k$. By the same approach, we may prove that $\eta_k(x)$ does not depend on k. Thus, (3.4) holds. □

By (3.4), (3.2) simply becomes $\eta e=P_k(\eta e)$, where $e=(1,1,\ldots,1)^T$ is a column vector of all 1's, and $P_k e=e$. Therefore, as a very special case, (3.2) holds for uni-chains.

3.2 Performance Potentials

3.2.1 Relative Performance Potentials

With confluencity, we can define the *relative performance potential*,[1] or the *relative potential* for short, for any two states $x, y\in\mathscr{S}_k$, $k=0,1,\ldots$, as

[1] It is called the *perturbation realization factor* in perturbation analysis (PA) (Cao 2007, 1994; Cao and Chen 1997); it measures the change in a total reward due to a shift (or a perturbation, in the terminology of PA) of the state from x to x'.

$$\gamma_k(x,y) := E\left\{ \sum_{l=k}^{k+\tau_k(x,y)} [f_l(X'_l) - f_l(X_l)] \Big| X'_k = y, X_k = x \right\}. \tag{3.6}$$

The strong confluencity (2.3) and Assumption 1.1b (1.10) imply[2]

$$|\gamma_k(x,y)| < L < \infty, \qquad \forall x, y \in \mathscr{S}_k, \ k = 0, 1, \ldots. \tag{3.7}$$

For any x, we have $\tau_k(x,x) = 0$ and $\gamma_k(x,x) = 0$. If $x \neq y$, then $\tau_k(x,y) > 0$, and

$$\begin{aligned}
\gamma_k(x,y) &= [f_k(y) - f_k(x)] \\
&+ E\left\{ \sum_{l=k+1}^{k+1+\tau_{k+1}(X_{k+1},Y_{k+1})} [f_l(X'_l) - f_l(X_l)] \Big| X'_k = y, X_k = x \right\} \\
&= [f_k(y) - f_k(x)] + E\left\{ \gamma_{k+1}(X_{k+1}, X'_{k+1}) \Big| X'_k = y, X_k = x \right\} \\
&= [f_k(y) - f_k(x)] + \sum_{x' \in \mathscr{S}_{k+1}} \sum_{y' \in \mathscr{S}_{k+1}} \gamma_{k+1}(x', y') P_k(x'|x) P_k(y'|y).
\end{aligned} \tag{3.8}$$

Let $\Gamma_k = [\gamma_k(x,y)]_{x,y \in \mathscr{S}_k}$, $e = (1, \ldots, 1)^T$, $f_k := (f_k(1), \ldots, f_k(S_k))^T$ and $F_k := e f_k^T - f_k e^T$. Then, (3.8) takes the form

$$\Gamma_k - P_k \Gamma_{k+1} P_k^T = F_k, \ \ k = 0, 1, \ldots. \tag{3.9}$$

By definition, we have

$$\gamma_k(x,x) = 0, \tag{3.10}$$

and

$$\gamma_k(x,y) = -\gamma_k(y,x). \tag{3.11}$$

Lemma 3.3 *If $|\gamma_k(x,y)| < \infty$ for all $x, y \in \mathscr{S}_k$, then the conservation law holds; i.e.,*

$$\gamma_k(x,y) + \gamma_k(y,z) + \gamma_k(z,x) = 0, \quad \forall x, y, z \in \mathscr{S}_k, \ \ k = 0, 1, \ldots. \tag{3.12}$$

Proof Equation (3.12) can be explained intuitively: By (3.6) and the Markov property, we have

$$\gamma_k(x,y) = E\left\{ \sum_{l=k}^{\infty} [f_l(X'_l) - f_l(X_l)] \Big| X'_k = y, X_k = x \right\}.$$

Then, (3.12) follows naturally. A rigorous proof is given in Sect. 3.6A. □

[2]Many results of this book may hold if we assume that (3.7), instead of (1.10), holds.

3.2.2 *Performance Potential, Poisson Equation, and Dynkin's Formula*

3.2.2.1 Performance Potential and Poisson Equation

By the conservation law (3.12), which follows the terminology for the potential energy in physics, there is a *performance potential function* $g_k(x)$, $x \in \mathscr{S}_k$, such that

$$\gamma_k(x, y) = g_k(y) - g_k(x), \quad k = 0, 1, \ldots. \tag{3.13}$$

Let $g_k = (g_k(1), \ldots, g_k(S_k))^T$, then, in (3.13) we have $\Gamma_k = e g_k^T - g_k e^T$, $e := (1, \ldots, 1)^T$, and from (3.9), we get

$$e[g_k - P_k g_{k+1} - f_k]^T = [g_k - P_k g_{k+1} - f_k] e^T,$$

i.e., $e[g_k - P_k g_{k+1} - f_k]^T$ is symmetric. Thus, there must be a constant c_k such that

$$g_k - P_k g_{k+1} - f_k = -c_k e, \quad k = 0, 1, \ldots; \tag{3.14}$$

this is the *Poisson equation*. c_k can be chosen as any real number and its choice does not affect the value of the relative potential $\gamma_k(x, y)$.

Lemma 3.4 *Any two potential vectors satisfying the same Poisson equation (3.14) corresponding to two different c_k's differ only by a constant vector $b_k e$, $k = 0, 1, \ldots$.*

Proof See Sect. 3.6B. □

For any fixed c_k, potentials g_k and g_{k+1} are unique only up to an additional constant; i.e., if g_k and g_{k+1} satisfy (3.14), then so do $g_k + ce$ and $g_{k+1} + ce$, for any real number c. This is similar to the potential energy in physics. The constant c_k determines the difference of the potential levels at time $k + 1$ and k; and adding any constant c to c_k at all levels k, $k = 0, 1, \ldots$, keeps the relative potentials $\gamma_k(x, y)$, $x, y \in \mathscr{S}$, $k = 0, 1, \ldots$ the same. This intrinsic property of potential provides flexibility in choosing the proper values of g_k.

Now, we choose

$$c_k = E[f_k(X_k)|X_0 = x_0], \quad k = 0, 1, \ldots. \tag{3.15}$$

Because we may add an arbitrary constant to g_k, $k = 0, 1, \ldots$, we may choose any initial state $X_0 = x_0$,[3] and set

[3] When we use "$X_0 = x_0$" as an initial state, we mean that if an equation holds for this particular initial state, then it may not hold if the initial state changes; e.g., in (3.15) we may have $c_k \neq E[f_k(X_k)|X_0 = x]$ for $x \neq x_0$; when we use "$X_0 = x$" as an initial state, we mean that the equation holds for all initial states $x \in \mathscr{S}_0$.

$$g_0(x_0) = 0,$$

and by (3.14), recursively we may prove that

$$E[g_k(X_k)|X_0 = x_0] = 0, \quad k = 0, 1, \ldots. \tag{3.16}$$

By (3.1) and (3.15), the average reward equals

$$\eta(x) = \eta(x_0) = \liminf_{K \to \infty} \frac{1}{K} \sum_{k=0}^{K-1} c_k. \tag{3.17}$$

3.2.2.2 The Form of $g(x)$

Iteratively, by (3.14), we have that for $k = 0, 1, \ldots,$

$$\begin{aligned} g_k &= P_k P_{k+1} g_{k+2} + P_k f_{k+1} + f_k - [c_k + c_{k+1}]e \\ &= \{\prod_{l=0}^{K-1} P_{k+l}\} g_{k+K} + \sum_{l=0}^{K-1} \{[\prod_{i=0}^{l-1} P_{k+i}] f_{k+l}\} - \{\sum_{l=0}^{K-1} c_{k+l}\} e, \end{aligned} \tag{3.18}$$

with $P_{k+i} := I$ for $i = -1$. Rewriting (3.18), we get that for any $y \in \mathscr{S}_k$,

$$\begin{aligned} g_k(y) &= E[g_K(X_K)|X_k = y] + \sum_{l=k}^{K-1} E[f_l(X_l)|X_k = y] - \sum_{l=k}^{K-1} c_l \\ &= \sum_{l=k}^{K-1} E[f_l(X_l) - c_l|X_k = y] + E[g_K(X_K)|X_k = y]. \end{aligned} \tag{3.19}$$

By (3.13), (3.16), and (3.7), at all time k, we have

$$|g_k(x)| < M < \infty, \quad \forall x \in \mathscr{S}_k, k = 0, 1, \ldots. \tag{3.20}$$

By (3.13), this requires $g_k(x)$ to be bounded for only one $x \in \mathscr{S}_k$, which is true because of (3.16).

Lemma 3.5 *With the strong confluencity (2.3), for any $y \in \mathscr{S}_k$, $k = 0, 1, \ldots,$ we have*

$$g_k(y) = \sum_{l=0}^{\infty} E[f_{k+l}(X_{k+l}) - c_{k+l}|X_k = y], \tag{3.21}$$

where c_k, $k = 0, 1, \ldots,$ are determined by (3.15), i.e., $c_k = E[f_k(X_k)|X_0 = x_0]$, $k = 0, 1, \ldots.$

Proof See Sect. 3.6C. □

3.2.2.3 Dynkin's Formula

Now, we define the discrete version of the *infinitesimal generator*, A_k, $k = 0, 1, \ldots$: For any sequence of vectors (or functions) $h_k(x)$, $x \in \mathscr{S}_k$, $k = 0, 1, \ldots$, define

$$(A_k h_k)(x) := \sum_{y \in \mathscr{S}_{k+1}} P_k(y|x) h_{k+1}(y) - h_k(x); \tag{3.22}$$

that is, the operator A_k acting on a sequence of vectors (or functions) $h_k(x)$, $x \in \mathscr{S}_k$, results in a sequence of vectors (or functions) denoted by $(A_k h_k)(x)$, or simply written as $A_k h_k(x)$. If action $\alpha = \alpha_k(x)$ is taken at state $x \in \mathscr{S}_k$, (3.22) is

$$(A_k^{\alpha_k(x)} h_k)(x) := \sum_{y \in \mathscr{S}_{k+1}} P_k^{\alpha_k(x)}(y|x) h_{k+1}(y) - h_k(x), \tag{3.23}$$

in which $A_k^{\alpha_k(x)}$ acting on a vector h_k results in a component of a vector. Under a decision rule α, (3.23) takes a vector form:

$$(A_k^{\alpha} h_k) := P_k^{\alpha} h_{k+1} - h_k.$$

For policy $u = \{\alpha_0^u, \alpha_1^u, \ldots\}$, at time k, the operator A_k^{α} can be written as $A_k^{\alpha_k^u}$, or simply as A_k^u. Similarly, we may write the transition matrix at k, $P_k^{\alpha_k^u}$, as P_k^u.

The Poisson equation (3.14) now takes the form

$$A_k g_k(x) + f_k(x) = c_k. \tag{3.24}$$

By the definition of the conditional expectation, we have

$$E\Big\{ -\Big[\sum_{y \in \mathscr{S}_{k+1}} P_k(y|X_k) h_{k+1}(y) \Big] + h_{k+1}(X_{k+1}) \Big| X_0 = x \Big\} = 0.$$

Thus, from the definition (3.22), we can easily verify that

$$E\{\sum_{k=0}^{K-1} [A_k h_k(X_k)] | X_0 = x\} = E[h_K(X_K)|X_0 = x] - h_0(x); \tag{3.25}$$

this is the discrete version of *Dynkin's formula*, and if the limit as $K \to \infty$ exists, then

$$E\{\sum_{k=0}^{\infty} [A_k h_k(X_k)] | X_0 = x\} = \lim_{K \to \infty} \{E[h_K(X_K)|X_0 = x]\} - h_0(x), \quad x \in \mathscr{S}_0.$$

3.3 Optimization of Average Rewards

In this section, with the performance potentials, we first derive the performance difference formula that compares the average rewards of any two policies and then obtain the average-reward optimality conditions. This approach is the relative optimization (cf. Cao 2007 for the case of THMCs).

3.3.1 The Performance Difference Formula

Consider any two independent Markov chains $\boldsymbol{X}$ and $\boldsymbol{X}'$ under any two policies denoted by $(\mathbb{P}, \mathscr{f})$ and $(\mathbb{P}', \mathscr{f}')$, respectively, with $\mathbb{P} = \{P_0, P_1, \ldots), \mathscr{f} = (f_0, f_1, \ldots)$ and $\mathbb{P}' = \{P'_0, P'_1, \ldots), \mathscr{f}' = (f'_0, f'_1, \ldots)$. Assume that $\mathscr{S}_k = \mathscr{S}'_k, k = 0, 1, \ldots$. Let $\eta, g_k, \{A_0, A_1, \ldots\}$ and $\eta', g'_k, \{A'_0, A'_1, \ldots\}$ be the quantities associated with these two Markov chains, respectively. Let $E := E^{\mathbb{P}}$ and $E' := E^{\mathbb{P}'}$ be the expectations corresponding to the measures generated by $\mathbb{P}$ and $\mathbb{P}'$, respectively.

Let $X_0 = X'_0 = x$. With c_k chosen as in (3.15), applying the Poisson equation (3.24) for $\boldsymbol{X}$ at any state on the sample path of $\boldsymbol{X}'$, $X'_k, k = 0, 1, \ldots$, we get

$$A_k g_k(X'_k) + f_k(X'_k) = c_k, \quad X'_k \in \mathscr{S}_k.$$

Taking the expectation on all the sample paths of $\boldsymbol{X}'$ and summing up yield

$$\sum_{k=0}^{K-1} E'\{(A_k g_k + f_k)(X'_k)|X'_0 = x\} = \sum_{k=0}^{K-1} c_k, \quad x \in \mathscr{S}_0.$$

Thus, from (3.17), we have

$$\eta = \liminf_{K\to\infty} \Big\{ \frac{1}{K} E'\{\sum_{k=0}^{K-1} (A_k g_k + f_k)(X'_k)|X'_0 = x\} \Big\}. \tag{3.26}$$

Now, applying Dynkin's formula (3.25) to $\boldsymbol{X}'$ with $h_k(x) = g_k(x)$ leads to

$$E'\{\sum_{k=0}^{K-1} [A'_k g_k(X'_k)]|X'_0 = x\} = E'[g_K(X'_K)|X'_0 = x] - g_0(x).$$

With the bound in (3.20), we have

$$\lim_{K\to\infty} \frac{1}{K} E'\{\sum_{k=0}^{K-1} [A'_k g_k(X'_k)]|X'_0 = x\} = 0.$$

By definition,

$$\eta' = \liminf_{K\to\infty} \frac{1}{K} E'\Big\{ \sum_{k=0}^{K-1} f'_k(X'_k) \Big| X'_0 = x \Big\}.$$

Adding the two sides of the above two equations together, and by Lemma 3.1, we have

$$\eta' = \liminf_{N\to\infty} \frac{1}{K} E'\Big\{ \sum_{k=0}^{K-1} (A'_k g_k + f'_k)(X'_k) \Big| X'_0 = x \Big\}. \tag{3.27}$$

Finally, from (3.26) and (3.27), we have the *performance difference formula*, as stated in the following lemma:

Lemma 3.6 *If Assumption 1.1 and the strong confluencity hold for two Markov chains* $\boldsymbol{X}$ *and* $\boldsymbol{X}'$ *with* $(\mathbb{P}, \boldsymbol{f})$ *and* $(\mathbb{P}', \boldsymbol{f}')$*, respectively, and* $X_0 = X'_0 = x$, $\mathscr{S}_k = \mathscr{S}'_k$, $k = 0, 1, \ldots$*; then,*

$$\begin{aligned}
&\eta' - \eta \\
&= \liminf_{K\to\infty} \frac{1}{K} E'\Big\{ \sum_{k=0}^{K-1} (A'_k g_k + f'_k)(X'_k) \Big| X'_0 = x \Big\} \\
&\quad - \liminf_{K\to\infty} \frac{1}{K} E'\Big\{ \sum_{k=0}^{K-1} (A_k g_k + f_k)(X'_k) \Big| X'_0 = x \Big\}.
\end{aligned} \tag{3.28}$$

3.3.2 Optimality Condition for Average Rewards

In this section, we provide a solution to the optimization problem formulated in Sect. 1.1. First, we define admissible policies.

Definition 3.2 A policy $u = (\mathbb{P}, \boldsymbol{f})$ is said to be *admissible*, if (a) Assumption 1.1 holds, and (b) $\boldsymbol{X}^u$ satisfies the strong confluencity.

The space of all admissible policies is denoted by $\mathscr{D}$. The goal is to find a policy in $\mathscr{D}$ that maximizes the average reward, as defined in (1.11), or (1.12).

Recall that the action taken at time k with $X_k = x \in \mathscr{S}_k$, $\alpha_k(x) \in \mathscr{A}_k(x)$, determines an operator $A_k^{\alpha_k(x)}$:

$$A_k^{\alpha_k(x)} h_k(x) = \sum_{y\in\mathscr{S}_{k,out}} P_k^{\alpha_k(x)}(y|x) h_{k+1}(y) - h_k(x) \ \ x \in \mathscr{S}_k.$$

For a policy $u = \{\alpha_0, \alpha_1, \ldots, \}$, we also write $A_k^u h_k(x) = A_k^{\alpha_k(x)} h_k(x)$.

Theorem 3.1 (Optimality condition) *Under Assumption 1.2, a policy* $u^* \in \mathscr{D}$ *is optimal, if*

$$(A_k^{u^*} g_k^{u^*} + f_k^{u^*})(y) = \max_{\alpha \in \mathscr{A}_k(y)} \{(A_k^{\alpha} g_k^{u^*} + f_k^{\alpha})(y)\} \tag{3.29}$$

holds for all $y \in \mathscr{S}_k$, $k = 0, 1, \ldots$; and η^{u^*} is the optimal performance.

Proof Replacing $(\mathbb{P}, f\!\!f)$ with $(\mathbb{P}^{u^*}, f\!\!f^{u^*})$ and $(\mathbb{P}', f\!\!f')$ with $(\mathbb{P}^u, f\!\!f^u)$ in the difference formula (3.28) yields

$$\begin{aligned}
&\eta^u - \eta^{u^*} \\
&= \liminf_{K \to \infty} \frac{1}{K} E^u \Big\{ \sum_{k=0}^{K-1} (A_k^u g_k^{u^*} + f_k^u)(X_k^u) \Big| X_0^u = x \Big\} \\
&\quad - \liminf_{K \to \infty} \frac{1}{K} E^u \Big\{ \sum_{k=0}^{K-1} (A_k^{u^*} g_k^{u^*} + f_k^{u^*})(X_k^u) \Big| X_0^u = x \Big\}.
\end{aligned} \tag{3.30}$$

If (3.29) holds for all $k = 0, 1, \ldots$, then

$$(A_k^u g_k^{u^*} + f_k^u)(y) \le (A_k^{u^*} g_k^{u^*} + f_k^{u^*})(y), \quad y \in \mathscr{S}_k,$$

for all $u \in \mathscr{D}$. Therefore,

$$\begin{aligned}
&\frac{1}{K} E^u \Big\{ \sum_{k=0}^{K-1} (A_k^u g_k^{u^*} + f_k^u)(X_k^u) \Big| X_0^u = x \Big\} \\
&\le \frac{1}{K} E^u \Big\{ \sum_{k=0}^{K-1} (A_k^{u^*} g_k^{u^*} + f_k^{u^*})(X_k^u) | X_0^u = x \Big\},
\end{aligned}$$

for all $K > 0$ and all $u \in \mathscr{D}$. Thus, we have $\eta^u \le \eta^{u^*}$, $\forall u \in \mathscr{D}$. Note that in (3.28), g_k corresponds to the c_k chosen in (3.15); however, because of Lemma 3.4, the optimality condition (3.29) holds for any form of g_k. □

Furthermore, the difference formula (3.28) contains all the information about the performance comparison in the entire time horizon $[0, \infty)$. (3.28) very naturally entails a weaker version of the optimality condition that takes care of the under-selectivity.

Theorem 3.2 (Optimality condition with under-selectivity) *Under Assumption 1.2, a policy $u^* \in \mathscr{D}$ in the admissible policy space $\mathscr{D}$ is optimal, if*

$$(A_k^{u^*} g_k^{u^*} + f_k^{u^*})(y) = \max_{\alpha \in \mathscr{A}_k(y)} \{(A_k^{\alpha} g_k^{u^*} + f_k^{\alpha})(y)\}, \quad y \in \mathscr{S}_k, \tag{3.31}$$

holds on every "frequently visited" subsequence of $k = 0, 1, \ldots$; or more precisely, if there exist a subsequence of $k = 0, 1, \ldots$, denoted by $k_0, k_1, \ldots, k_l, \ldots$, a sequence of states visited by $\boldsymbol{X}^{u^*}$ at $k_0, k_1, \ldots$, denoted by $x_{k_0}, x_{k_1}, \ldots, x_{k_l}, \ldots$, $x_{k_l} \in \mathscr{S}_{k_l}$, $l =$

0, 1, ..., and a sequence of actions $\alpha_{k_0}, \alpha_{k_1}, \ldots, \alpha_{k_l} \in \mathscr{A}_{k_l}(x_{k_l})$, $l = 0, 1, \ldots$, such that (3.31) does not hold on k_l, $l = 0, 1, \ldots$, i.e., $(A_{k_l}^{u^*} g_{k_l}^{u^*} + f_{k_l}^{u^*})(x_{k_l}) < (A_{k_l}^{\alpha_{k_l}} g_{k_l}^{u^*} + f_{k_l}^{\alpha_{k_l}})(x_{k_l})$, then we must have

$$\lim_{n\to\infty} \frac{1}{k_n} \sum_{l=1}^{n} \left\{ (A_{k_l}^{u^*} g_{k_l}^{u^*} + f_{k_l}^{u^*})(x_{k_l}) - (A_{k_l}^{\alpha_{k_l}} g_{k_l}^{u^*} + f_{k_l}^{\alpha_{k_l}})(x_{k_l}) \right\} = 0. \tag{3.32}$$

Such a sequence is called a "non-frequently visited" sequence.

Proof By Theorem 3.1, if (3.31) holds for all $k = 0, 1, \ldots$, then by the difference formula (3.30), $\eta^u \le \eta^{u^*}$ for all $u \in \mathscr{D}$, and u^* is optimal. Furthermore, if (3.31) does not hold for a policy, denoted by u, at some time instants denoted by $k_0, k_1, \ldots$, then we can prove that

$$\lim_{n\to\infty} \frac{1}{k_n} E^u \Big\{ \sum_{l=1}^{n} \Big[(A_{k_l}^{u^*} g_{k_l}^{u^*} + f_{k_l}^{u^*})(X_{k_l}^u) - (A_{k_l}^{u} g_{k_l}^{u^*} + f_{k_l}^{u})(X_{k_l}^u) \Big] \Big| X_0^u = x \Big\} = 0. \tag{3.33}$$

To this end, we first observe that at any k_l there must exist a state x_{k_l} with action α_{k_l} used in policy u such that

$$\begin{aligned} &(A_{k_l}^{u^*} g_{k_l}^{u^*} + f_{k_l}^{u^*})(x_{k_l}) - (A_{k_l}^{\alpha_{k_l}} g_{k_l}^{u^*} + f_{k_l}^{\alpha_{k_l}})(x_{k_l}) \\ &\le E^u \Big\{ \Big[(A_{k_l}^{u^*} g_{k_l}^{u^*} + f_{k_l}^{u^*})(X_{k_l}^u) - (A_{k_l}^{u} g_{k_l}^{u^*} + f_{k_l}^{u})(X_{k_l}^u) \Big] \Big| X_0^u = x \Big\}. \end{aligned}$$

Thus, if the limit in (3.33) is less than zero, then we must have

$$\lim_{n\to\infty} \frac{1}{k_n} \sum_{l=1}^{n} \left\{ (A_{k_l}^{u^*} g_{k_l}^{u^*} + f_{k_l}^{u^*})(x_{k_l}) - (A_{k_l}^{\alpha_{k_l}} g_{k_l}^{u^*} + f_{k_l}^{\alpha_{k_l}})(x_{k_l}) \right\} < 0,$$

which contradicts to (3.32). Thus, (3.33) holds. This cannot change the inequality $\eta^u \le \eta^{u^*}$, (cf. (3.30)), and u^* is optimal. □

For the necessary condition, we need some uniformity conditions.

Definition 3.3 We say that

(a) A sequence a_n, $n = 0, 1, \ldots$, is *uniformly positive*, if there is an $\varepsilon > 0$ such that $a_n > \varepsilon$, for all $n = 0, 1, \ldots$;
(b) $a_n > b_n$ uniformly, if the sequence $a_n - b_n$ is uniformly positive; and
(c) A set of real numbers A is *uniformly positive*, or $a > 0$ uniformly on A, if there is an $\varepsilon > 0$ such that $a > \varepsilon$ for all $a \in A$.

With this definition, we may add the following statement to Theorem 3.3:

"If in the subsequence, $(A_{k_l}^{\alpha_{k_l}} g_{k_l}^{u^*} + f_{k_l}^{\alpha_{k_l}})(x_{k_l}) > (A_{k_l}^{u^*} g_{k_l}^{u^*} + f_{k_l}^{u^*})(x_{k_l})$ uniformly on $k_0, k_1, \ldots$, then

$$\lim_{n\to\infty} \frac{n}{k_n} = 0$$

holds." This explains the words "non-frequently visited sequence."

In the rest of the section, without loss of generality, we assume there is no new state joining the process at $k > 0$.[4] In this case, it holds that

$$\mathscr{S}_{k-1,out} = \mathscr{S}_k, \ \ k = 1, 2, \ldots. \tag{3.34}$$

Because of the confluencity, when l is large enough, $\mathscr{P}(X_{k+l} = y|X_k = x) > 0$, for all $y \in \mathscr{S}_{k+l}$. (By Assumption 1.2, $\mathscr{S}_k, k = 0, 1, \ldots$, are the same for all policies.)

Definition 3.4 A TNHMC $\boldsymbol{X} = \{X_k, k \geq 0\}$ is said to be *weakly recurrent*, if there is an $\varepsilon > 0$ such that if $\mathscr{P}(X_{k+l} = y|X_k = x) > 0$, then

$$\mathscr{P}(X_{k+l} = y|X_k = x) > \varepsilon, \tag{3.35}$$

for any $x \in \mathscr{S}_k$ and all $y \in \mathscr{S}_{k+l}$, and $k, l = 0, 1, \ldots$.

The definition is equivalent to: if $\mathscr{P}(X_{k+1} = y|X_k = x) > 0$, then $\mathscr{P}(X_{k+1} = y|X_k = x) > \varepsilon$, for all $k = 0, 1, \ldots$.

Assumption 3.3 $\boldsymbol{X}^u$ is weakly recurrent for all policies $u \in \mathscr{D}$.

This assumption does not require the ε in (3.35) to be the same for all $u \in \mathscr{D}$.

Theorem 3.3 (Necessary optimality condition) *If Assumptions 1.2 and 3.3 hold and* $\{x_{k_l} \in \mathscr{S}_{k_l}, \ l = 0, 1, \ldots\}$ *is the subsequence in Theorem 3.2, then condition (3.32) (assuming the limit exists) is necessary for* u^* *to be optimal.*

Proof Suppose (3.32) does not hold for a subsequence $k_0, k_1, \ldots$, with $(A_{k_l}^{u^*} g_{k_l}^{u^*} + f_{k_l}^{u^*})(x_{k_l}) < (A_{k_l}^{\alpha_{k_l}} g_{k_l}^{u^*} + f_{k_l}^{\alpha_{k_l}})(x_{k_l})$; that is, instead of (3.32), we have

$$\lim_{n\to\infty} \frac{1}{k_n} \sum_{l=1}^{n} \Big\{(A_{k_l}^{u^*} g_{k_l}^{u^*} + f_{k_l}^{u^*})(x_{k_l}) - (A_{k_l}^{\alpha_{k_l}} g_{k_l}^{u^*} + f_{k_l}^{\alpha_{k_l}})(x_{k_l})\Big\} < 0, \tag{3.36}$$

then we can prove that u^* is not optimal. To this end, we construct a new policy, denoted by $\tilde{u} := (\tilde{\mathbb{P}}, \tilde{\boldsymbol{f}}) = \{\tilde{P}_0, \tilde{f}_0; \tilde{P}_1, \tilde{f}_1; \ldots\}$ such that $\tilde{P}_k = P_k^{u^*}$ and $\tilde{f}_k = f_k^{u^*}$, for all $k \notin \{k_0, k_1, \ldots\}$; $\tilde{P}_{k_l}(y|x) = P_{k_l}^{u^*}(y|x)$, $\tilde{f}_{k_l}(x) = f_{k_l}^{u^*}(x)$ for all $x \neq x_{k_l}$, $l = 0, 1, \ldots$; and

$$\tilde{P}(y|x_{k_l}) := P^{\alpha_{k_l}}(y|x_{k_l}), \ \ \tilde{f}_{k_l}(x_{k_l}) = f_{k_l}^{\alpha_{k_l}}(x_{k_l}), \ l = 0, 1, \ldots.$$

[4]Otherwise, the notational complexity increases, see Cao [2015].

With the policy $\tilde{u}$ defined above, we have

$$(A_{k_l}^{u^*} g_{k_l}^{u^*} + f_{k_l}^{u^*})(x_{k_l}) - (A_{k_l}^{\tilde{u}} g_{k_l}^{u^*} + f_{k_l}^{\tilde{u}})(x_{k_l}) < 0,$$

$l = 0, 1, \ldots$, and

$$(A_k^{u^*} g_k^{u^*} + f_k^{u^*})(x) = (A_k^{\tilde{u}} g_k^{u^*} + f_k^{\tilde{u}})(x),$$
$$for\ k \notin \{k_0, k_1, \ldots, \},\ and\ x \neq x_{k_l},\ if\ k \in \{k_0, k_1, \ldots\}.$$

It is clear that

$$\lim_{K \to \infty} \frac{1}{K} E^{\tilde{u}} \Big\{ \sum_{k=0}^{K-1} [(A_k^{u^*} g_k^{u^*} + f_k^{u^*}) - (A_k^{\tilde{u}} g_k^{u^*} + f_k^{\tilde{u}})](X_k^{\tilde{u}}) \Big| X_0^{\tilde{u}} = x \Big\}$$
$$\leq \lim_{n \to \infty} \frac{1}{k_n} E^{\tilde{u}} \Big\{ \sum_{l=0}^{n} [(A_{k_l}^{u^*} g_{k_l}^{u^*} + f_{k_l}^{u^*}) - (A_{k_l}^{\tilde{u}} g_{k_l}^{u^*} + f_{k_l}^{\tilde{u}})](X_{k_l}^{\tilde{u}}) \Big| X_0^{\tilde{u}} = x \Big\}.$$

Furthermore, by Assumption 3.3 and (3.35), for any policy $\tilde{u}$ and all k_l, we have

$$E^{\tilde{u}} \Big\{ (A_{k_l}^{u^*} g_{k_l}^{u^*} + f_{k_l}^{u^*})(X_{k_l}) - (A_{k_l}^{\tilde{u}} g_{k_l}^{u^*} + f_{k_l}^{\tilde{u}})(X_{k_l}) | X_0^{\tilde{u}} = x \Big\}$$
$$\leq \varepsilon \Big\{ (A_{k_l}^{u^*} g_{k_l}^{u^*} + f_{k_l}^{u^*})(x_{k_l}) - (A_{k_l}^{\tilde{u}} g_{k_l}^{u^*} + f_{k_l}^{\tilde{u}})(x_{k_l}) \Big\} \ (< 0).$$

From (3.36) and the above two equations, we conclude that

$$\lim_{K \to \infty} \frac{1}{K} E^{\tilde{u}} \Big\{ \sum_{k=0}^{K-1} [(A_k^{u^*} g_k^{u^*} + f_k^{u^*}) - (A_k^{\tilde{u}} g_k^{u^*} + f_k^{\tilde{u}})](X_k^{\tilde{u}}) \Big| X_0^{\tilde{u}} = x \Big\} < 0.$$

This leads to $\eta^{\tilde{u}} > \eta^{u^*}$ (cf. (3.30)); i.e., u^* is not optimal. □

Finally, if the "sup" in (1.11) cannot be reached, we need to find the value of $\sup_{u \in \mathscr{D}} \eta^u$ in (1.12).

Theorem 3.4 *Under Assumption 1.2, if there is a sequence of functions $h_k(x)$, $k = 0, 1, \ldots$, $x \in \mathscr{S}_k$, and a sequence of real numbers c_k, $k = 0, 1, \ldots$, such that*

$$\lim_{K \to \infty} \frac{1}{K} \{E[h_K(X_K^u) | X_0^u = x_0] = 0$$

for all $u \in \mathscr{D}$, and

$$\sup_{\alpha \in \mathscr{A}_k(x)} \{(A_k^\alpha h_k + f_k^\alpha)(x)\} = c_k,\ \forall x \in \mathscr{S}_k,\ k = 0, 1, \ldots, \tag{3.37}$$

then

$$\sup_{u\in\mathscr{D}}\{\eta^u\} = \liminf_{K\to\infty} \frac{1}{K}\sum_{k=0}^{K-1} c_k. \tag{3.38}$$

Proof For any policy $u \in \mathscr{D}$, by Dynkin's formula, we have

$$E\{\sum_{k=0}^{K-1}[A_k^u h_k(X_k^u)]|X_0^u = x_0\} = E[h_K(X_K^u)|X_0^u = x_0] - h_0(x_0).$$

From (3.37), we get $A_k^u h_k(x) \le c_k - f_k^u(x)$ for all $x \in \mathscr{S}_k$ and $k = 0, 1, \ldots$, and thus

$$E\{\sum_{k=0}^{K-1}[c_k - f_k^u(X_k^u)]|X_0^u = x_0\} \ge E[h_K(X_K^u)|X_0^u = x_0] - h_0(x_0);$$

that is

$$\frac{1}{K}E\{\sum_{k=0}^{K-1} f_k^u(X_k^u)|X_0^u = x_0\} \le \frac{1}{K}\sum_{k=0}^{K-1} c_k - \frac{1}{K}\{E[h_K(X_K^u)|X_0^u = x_0] - h_0(x_0)\}. \tag{3.39}$$

Letting $K \to \infty$, we obtain

$$\eta^u := \liminf_{K\to\infty} \frac{1}{K}E\{\sum_{k=0}^{K} f_k^u(X_k^u)|X_0^u = x_0\} \le \liminf_{K\to\infty} \frac{1}{K}\sum_{k=0}^{K} c_k. \tag{3.40}$$

Furthermore, because the actions in (3.37) at different time k can be chosen independently, for any sequence of positive numbers $\varepsilon_0, \varepsilon_1, \ldots$, with $\lim_{l\to\infty} \varepsilon_l = 0$, there exists a sequence of policies $u_0, u_1, \ldots$, with $u_l := (\mathbb{P}_l, \mathbb{f}_l)$, $\mathbb{P}_l = \{P_{l,0}, P_{l,1}, \ldots\}$, $\mathbb{f}_l = \{f_{l,0}, f_{l,1}, \ldots\}$, such that for every $u_l, l = 1, 2 \ldots$, we have

$$(A_k^{u_l} h_k + f_k^{u_l})(x) > c_k - \varepsilon_l, \quad for\ all\ k = 0, 1, \ldots.$$

Similar to (3.39), we have

$$
\begin{aligned}
\eta^{u_l} &:= \liminf_{K\to\infty} \frac{1}{K} E\{\sum_{k=0}^{K-1} f^{u_l}(k, X_k^{u_l})|X_0^{u_l} = x_0\} \\
&> \liminf_{K\to\infty} \frac{1}{K} \sum_{k=0}^{K-1} c_k - \varepsilon_l,
\end{aligned}
$$

which, in conjunction with (3.40), leads to (3.38). □

As shown in Theorem 3.1, if the value of "sup" in (3.37) can be reached by an optimal policy u^* with $\alpha = u_k^*(x)$, then in Theorem 3.4, $h_k(x) = g_k^{u^*}(x)$ is the potential function of the optimal policy.

In summary, the main optimality equation is (3.31), which is similar to that of THMCs (Cao 2007; Hernández-Lerma and Lasserre 1996; Puterman 1994). Computation procedures similar to those for THMCs, such as value iteration, policy iteration, perturbation analysis, and reinforcement learning, may be developed. However, because of the infinitely many terms of g_k's, the algorithms may be more complicated; these will be left for further research.

3.4 Bias Optimality

The long-run average reward is an under-selective criterion that does not depend on the decision rules in any finite period, and a long-run average optimal policy may behave poorly in the initial period. In this section, we address this issue by analyzing the transient behavior of the average-reward optimal policies. Among all these optimal policies, we may find one that optimizes a transient performance called bias; optimality equations for this bias optimal policy are provided.

3.4.1 Problem Formulation

3.4.1.1 The Transient Performance (bias)

Since the infinite sum $\sum_{l=k}^{\infty} E[f_l(X_l)|X_k = x]$ may not exist, we define the bias as

$$
\hat{g}_k(x) = E\left\{ \sum_{l=k}^{\infty} [f_l(X_l) - \eta] \middle| X_k = x \right\}, \tag{3.41}
$$

with η being the average reward as the limit defined in (1.7). Let $\hat{g}_k = (\hat{g}_k(1), \dots \hat{g}_k(S_k))^T$, $k = 0, 1, \dots$.[5] The bias represents the *transient performance*.

Assumption 3.4 $\hat{g}_k(x)$ exists, i.e., (3.41) converges for all $x \in \mathscr{S}_k$ and $k = 0, 1, \dots$.

Lemma 3.7 *Under Assumption 3.4,*

(a) $\lim_{K\to\infty} E[f_K(X_K)|X_0 = x] = \eta$, *and the limit in (1.7) exists,*
(b) $\lim_{K\to\infty} E[\hat{g}_K(X_K)|X_k = x] = 0$, *for all* $x \in \mathscr{S}_k$, $k = 0, 1, \dots$,
(c) $E[\hat{g}_k(X_k)|X_0 = x]$ *is bounded for all* $k = 0, 1, \dots$, *and*
(d) If $\boldsymbol{X}$ *is weakly recurrent, then* $\hat{g}_k(x)$ *is bounded for all* $x \in \mathscr{S}_k$, *and* $k = 0, 1, \dots$.

Proof (a) is necessary for (3.41) to exist. From (3.41), we have

$$\hat{g}_k(x) = E\left\{ \sum_{l=k}^{K-1} [f_l(X_l) - \eta] \Big| X_k = x \right\} + E[\hat{g}_K(X_K)|X_k = x].$$

Letting $K \to \infty$ yields (b). (c) follows from (b), and (d) follows from (c) and (3.35). □

It is easy to verify that $\hat{\gamma}_k(x, y) := \hat{g}_k(y) - \hat{g}_k(x)$, $k = 0, 1, \dots$, satisfy (3.9) and therefore are the relative performance potentials and that $\hat{g}_k(x)$, $x \in \mathscr{S}_k$, $k = 0, 1, \dots$, satisfy the Poisson equation (3.14) with $\hat{c}_k = \eta$:

$$\hat{g}_k - P_k \hat{g}_{k+1} - f_k = -\eta e, \quad k = 0, 1, \dots. \tag{3.42}$$

3.4.1.2 The Space of Average Optimal Policies and Bias Optimality

Our goal is to optimize the bias while keeping the average-reward optimal. Thus, we need to find the space of all the average-reward optimal policies. Again, we assume there are no new states added to the chain at any time, i.e., (3.34) holds. It follows from the definitions (1.3) and (1.4) that for any k, k', $k < k'$, and any state $y \in \mathscr{S}_{k'}$, there is a state $x \in \mathscr{S}_k$ such that

$$\mathscr{P}[X_{k'} = y|X_k = x] > 0. \tag{3.43}$$

Lemma 3.8 *Let* $u^* = (\mathbb{P}^*, f^*)$ *be an average-reward optimal policy,* $g_k^{u^*}$ *be its performance potential, and Assumption 3.3 hold. Any other admissible policy* $u = (\mathbb{P}, f)$ *is average-reward optimal if and only if*

$$(A_k^{u^*} g_k^{u^*} + f_k^{u^*})(x) = (A_k^u g_k^{u^*} + f_k^u)(x), \quad x \in \mathscr{S}_k, \tag{3.44}$$

[5] We use g_k to denote the performance potential, and $\hat{g}_k$ to denote the bias, which is a special form of potentials as shown in (3.41). g_k and $\hat{g}_k$ differ by a constant c_k, $k = 0, 1, \dots$.

holds on every "frequently visited" subsequence of $k = 0, 1, \ldots$. *More precisely, if there exist a subsequence of* $k = 0, 1, \ldots$, *denoted by* $k_0, k_1, \ldots, k_l, \ldots$, *and a sequence of states, denoted by* $x_{k_0}, x_{k_1}, \ldots, x_{k_l}, \ldots$, $x_{k_l} \in \mathscr{S}_{k_l}$, $l = 0, 1, \ldots$, *such that (3.44) does not hold on* k_l, $l = 0, 1, \ldots$, *i.e.,* $(A^{u^*}_{k_l} g^{u^*}_{k_l} + f^{u^*}_{k_l})(x_{k_l}) < (A^{u}_{k_l} g^{u^*}_{k_l} + f^{u}_{k_l})(x_{k_l})$, *then we must have*

$$\lim_{n\to\infty} \frac{1}{k_n} \sum_{l=1}^{n} \Big\{ (A^{u^*}_{k_l} g^{u^*}_{k_l} + f^{u^*}_{k_l})(x_{k_l}) - (A^{u}_{k_l} g^{u^*}_{k_l} + f^{u}_{k_l})(x_{k_l}) \Big\} = 0.$$

In addition, if in the subsequence, $(A^{\alpha_{k_l}}_{k_l} g^{u^*}_{k_l} + f^{\alpha_{k_l}}_{k_l})(x_{k_l}) > (A^{u^*}_{k_l} g^{u^*}_{k_l} + f^{u^*}_{k_l})(x_{k_l})$, $\alpha_k = \alpha^u_k(x_k)$, *uniformly, then we must have*

$$\lim_{n\to\infty} \frac{n}{k_n} = 0. \tag{3.45}$$

Proof The "if" part follows directly from the performance difference formula (3.28). The "only if" part can be proved by constructing a policy $\tilde{u}$ such that $\eta^{\tilde{u}} < \eta^*$ (cf. Theorem 3.2). □

From (3.44), we may define the set of all "average-reward optimal actions" as

$$\mathscr{A}_{0,k}(x) := \Big\{ \alpha \in \mathscr{A}_k(x) : \sum_y P^{\alpha}_k(y|x) g^{u^*}_{k+1}(y) + f^{\alpha}_k(x) = \sum_y P^{u^*}_k(y|x) g^{u^*}_{k+1}(y) + f^{u^*}_k(x) \Big\}, \quad x \in \mathscr{S}_k, \tag{3.46}$$

where u^* is called a *reference policy*. The product space of the action sets for all $x \in \mathscr{S}_k$ is the space of all the decision rules:

$$\mathscr{C}_{0,k} := \prod_{x\in\mathscr{S}_k} \mathscr{A}_{0,k}(x), \tag{3.47}$$

from which we get the policy space

$$\mathscr{C}_0 := \prod_{k=0}^{\infty} \mathscr{C}_{0,k}, \tag{3.48}$$

with $\prod$ denoting the Cartesian product.

Let the *space of all average-reward optimal policies* be $\mathscr{D}_0$. By Lemma 3.8, a policy in $\mathscr{D}_0$ differs from one in $\mathscr{C}_0$ only by a "non-frequently visited" sequence of time instants and $\mathscr{C}_0 \subset \mathscr{D}_0$.

By Lemma 3.4, for any two potentials g'_k and g_k, $k = 0, 1, \ldots$, satisfying the same Poisson equation, we have $g'_k - g_k = b_k e$; thus, for any transition probability matrix P_k, we have $P_k g'_{k+1} = P_k g_{k+1} + b_{k+1} e$. Therefore, by (3.46), different versions of potentials lead to the same set of optimal policies and the same space $\mathscr{C}_0$. Furthermore, we have

Lemma 3.9 *(a) For any policy $u \in \mathscr{C}_0$, with $\mathscr{C}_0$ defined in (3.46)–(3.48), we have*

$$g_k^u - g_k^{u^*} = b_k e, \quad k = 0, 1, \ldots, \tag{3.49}$$

where b_k, $k = 0, 1, \ldots$, are constants, depending on u and u^*.
(b) The set $\mathscr{C}_0$ is independent of the choices of u^* in $\mathscr{C}_0$.

Proof (a) This follows directly from Lemma 3.4, because $g_k^{u^*}$ and g_k^u, $k = 0, 1, \ldots$, satisfy the same Poisson equation, i.e., $A_k^u g_k^{u^*} + f_k^u = \eta^* e$ (cf. (3.46)), and $A_k^u g_k^u + f_k^u = c_k e$, with different constants η^* and c_k.

(b) Let $u \in \mathscr{C}_0$ be any policy; then by (3.49), we have $A_k^{u'} g_k^u = A_k^{u'} g_k^{u^*}$ for any policy u. Thus, by (3.46)–(3.48), the set $\mathscr{C}_0$ is the same if u^* is replaced by u. □

By this lemma, if $(\mathbb{P}, f\!\!\!f)$ and $(\mathbb{P}', f\!\!\!f')$ are any two policies in $\mathscr{D}_0$, we have

$$P_k g'_{k+1} + f_k = P'_k g'_{k+1} + f'_k, \quad x \in \mathscr{S}_k, \quad k = 0, 1, \ldots, \tag{3.50}$$

holds except at a non-frequently visited sequence $k_0, k_1, \ldots, k_l, \ldots$.

However, $\mathscr{C}_0$ is not unique, because the policy u^* in (3.46) may differ at a non-frequently visited sequence. By Lemma 3.9(b), from now on, we choose a policy $u^* \in \mathscr{C}_0$ such that (3.29) holds, i.e., $(A_k^{u^*} g_k^{u^*} + f_k^{u^*})(y) = \max_{\alpha \in \mathscr{A}_k(y)}\{(A_k^\alpha g_k^{u^*} + f_k^\alpha)(y)\}$, for all $y \in \mathscr{S}_k$, and $k = 0, 1, \ldots$. With this choice, by Lemma 3.9(a), the following equation holds for any $u \in \mathscr{C}_0$:

$$(A_k^u g_k^u + f_k^u)(y) = \max_{\alpha \in \mathscr{A}_k(y)} \{(A_k^\alpha g_k^u + f_k^\alpha)(y)\}, \quad y \in \mathscr{S}_k, \; k = 0, 1, \ldots. \tag{3.51}$$

The goal is to optimize the bias in $\mathscr{D}_0$; i.e., to find a $u^* \in \mathscr{D}_0$ such that

$$\hat{g}_k^{u^*}(x) = \max_{u \in \mathscr{D}_0}\{\hat{g}_k^u(x)\},$$

for all $x \in \mathscr{S}_k$, $k = 0, 1, \ldots$. We will see that under some conditions, such a policy exists. We follow the same procedure as that for the average optimal policies, with small modifications.

3.4.2 Bias Optimality Conditions

3.4.2.1 Potential of Potentials (Bias of Bias)

Similar to (3.6), we define the *relative bias potentials* for any $x, y \in \mathscr{S}_k$ as[6]:

$$\chi_k(x, y) := E\Big\{ \sum_{l=k}^{k+\tau_k(x,y)} [\hat{g}_l(X_l) - \hat{g}_l(X'_l)] \Big| X'_k = y, X_k = x \Big\}, \quad k = 0, 1, \ldots, \tag{3.52}$$

where X_l and X'_l, $l = 0, 1, \ldots$, are two independent sample paths under the same policy with two different initial states x and y, respectively. Equations similar to (3.9)–(3.12) hold for $\chi_k(x, y)$, $x, y \in \mathscr{S}_k$, $k = 0, 1, \ldots$. Thus, we can define the *bias potentials* (or *potentials of potential*) $w_k(x)$, $x \in \mathscr{S}_k$, as

$$\chi_k(x, y) = w_k(y) - w_k(x), \quad k = 0, 1, \ldots, \quad k = 0, 1, \ldots. \tag{3.53}$$

Set the vector $w_k = (w_k(1), \ldots, w_k(S_k))^T$; then, we have (cf. (3.14)) $w_k - P_k w_{k+1} + \hat{g}_k = -a_k e, \quad k = 0, 1, \ldots$. The constants a_k can be chosen arbitrarily, and we set $a_k = 0, k = 0, 1, \ldots$ (cf. (3.15) and (3.16))), so

$$w_k - P_k w_{k+1} + \hat{g}_k = 0, \quad k = 0, 1, \ldots. \tag{3.54}$$

This is the *Poisson equation for bias potentials*; with the operator A_k defined in (3.22), we have

$$A_k w_k(x) = \hat{g}_k(x). \tag{3.55}$$

By (3.54), we have (cf. (3.18)) $w_k = \{\prod_{l=k}^{K-1} P_l\} w_K - \sum_{l=k}^{K-1} \{[\prod_{i=k}^{l-1} P_i] \hat{g}_l\}$, with $\prod_{l=k}^{k-1} P_l = I$. Rewriting, we get that

$$w_k(y) = E[w_K(X_K) | X_k = y] - \sum_{l=k}^{K-1} E[\hat{g}_l(X_l) | X_k = y],$$

for any $y \in \mathscr{S}_k$ and $K > k$ (cf. (3.19)). Therefore,

$$\frac{1}{K} E[w_K, (X_K) | X_k = x] = \frac{1}{K} w_k(x) - \frac{1}{K} \sum_{l=k}^{K-1} E[\hat{g}_l(X_l) | X_k = x].$$

By Lemma 3.7.b, we get

[6] Note that $\chi_k(x, y)$ in (3.52) corresponds to the negative value of $\gamma(x, y)$ in (3.6). In this way, the bias Poisson equation takes the form of (3.54) (cf. (3.24)), and the bias difference formula (3.60) takes a similar form as the performance difference formula (3.28).

$$\lim_{K\to\infty} \frac{1}{K} E[w_K(X_K)|X_k = x] = 0. \tag{3.56}$$

By the strong confluencity and (3.7), we have

$$\chi_k(x, y) < L < \infty, \qquad \forall x, y \in \mathscr{S}_k, \ k = 0, 1, \ldots. \tag{3.57}$$

From (3.56), we can prove the following lemma:

Lemma 3.10 *Let $\boldsymbol{X}$ and $\boldsymbol{X}'$ be two TNHMCs with two different policies u and u'. Under Assumption 3.4 and with the strong confluencity, we have*

$$\lim_{K\to\infty} \frac{1}{K} E'[w_K(X'_K)]|X'_k = x] = 0. \tag{3.58}$$

Proof By (3.57), we have $|w_k(x) - w_k(y)| < M$ for all $x, y \in \mathscr{S}_k$ and $k \geq 0$, for some finite M. Choose any $y_k \in \mathscr{S}_k$, then for all $y \in \mathscr{S}_k$, we have $w_k(y_k) - M < w_k(y) < w_k(y_k) + M, \ y \in \mathscr{S}_k$. Therefore,

$$w_k(y_k) - M < E'[w_k(X'_k)]|X'_k = x] < w_k(y_k) + M, \tag{3.59}$$

and

$$w_k(y_k) - M < E[w_k(X_k)]|X_k = x] < w_k(y_k) + M.$$

Now, from (3.56), we have

$$\lim_{K\to\infty} \frac{1}{K} w_K(y_K) = \lim_{K\to\infty} \frac{1}{K} E[w_K(X_K)|X_0 = x] = 0;$$

and then from (3.59), we get (3.58):

$$\lim_{K\to\infty} \frac{1}{K} E'[w_K(X'_K)|X'_0 = x] = \lim_{K\to\infty} \frac{1}{K} w_K(y_K) = 0. \qquad \square$$

3.4.2.2 The Bias Difference Formula

Consider two Markov chains $\boldsymbol{X} = \{X_k, k = 0, 1, \ldots\}$ and $\boldsymbol{X}' = \{X'_k, k = 0, 1, \ldots\}$, $X_0 = x'_0 = x$, under two admissible policies $u = (\mathbb{P}, \boldsymbol{f})$ and $u' = (\mathbb{P}', \boldsymbol{f}'), u, u' \in \mathscr{D}$; let η and η', $\hat{g}_k$ and $\hat{g}'_k$, and w_k and w'_k, be their average rewards, potentials, and bias potentials, respectively.

Lemma 3.11 *Suppose Assumption 3.4 holds. For two admissible policies $u = (\mathbb{P}, \boldsymbol{f})$, $u' = (\mathbb{P}', \boldsymbol{f}') \in \mathscr{D}$, if $\eta = \eta'$, then*

$$\begin{aligned}&\hat{g}'_0(x) - \hat{g}_0(x)\\&= \lim_{K\to\infty}\frac{1}{K}\sum_{k=0}^{K-1} E'\{[A'_k w_k(X'_k) - A_k w_k(X'_k)]\Big|X'_0 = x\}\\&+ \lim_{K\to\infty}\frac{1}{K}\sum_{l=1}^{K-1}(K-l-1)\Big[E'\Big\{[(P'_{l-1}\hat{g}_l + f'_{l-1})(X'_{l-1})\\&- (P_{l-1}\hat{g}_l + f_{l-1})(X'_{l-1})]\Big|X'_0 = x\Big\}\Big].\end{aligned} \tag{3.60}$$

Proof See Sect. 3.6D. □

If $(\mathbb{P}, \mathbb{f}), (\mathbb{P}', \mathbb{f}') \in \mathscr{C}_0$, as defined in (3.48), then by (3.50), the second term in (3.60) is zero and it becomes

$$\begin{aligned}&\hat{g}'_0(x) - \hat{g}_0(x)\\&= \lim_{K\to\infty}\frac{1}{K}\sum_{k=0}^{K-1} E'\Big\{[A'_k w_k(X'_k) - A_k w_k(X'_k)]\Big|X'_0 = x\Big\}.\end{aligned} \tag{3.61}$$

However, if $(\mathbb{P}, \mathbb{f}), (\mathbb{P}', \mathbb{f}') \in \mathscr{D}_0 - \mathscr{C}_0$, the second term in (3.60) may not be zero, since the items in the second summation $\sum_{l=1}^{K-1}$ may not be zero on a "non-frequently visited" sequence of time. These nonzero terms do not affect the value of the long-run average, but they do affect the transient value $\hat{g}'_0(y)$, because of the factor $(K-l-1)$ in the second term of (3.60).

3.4.2.3 Bias Optimality Conditions

Again, to derive the optimality conditions, we require some uniformity. First, for an optimal policy u^*, we set

$$\Delta_k(\alpha, x) := (P_k^\alpha \hat{g}_{k+1}^{u^*} + f_k^\alpha)(x) - (P_k^{u^*}\hat{g}_{k+1}^{u^*} + f_k^{u^*})(x). \tag{3.62}$$

By (3.49), this value does not depend on the choice of u^*. Set

$$\Lambda_k := \{\Delta_k(\alpha, x) > 0 : \ all\ \alpha \in \mathscr{A}_k(x),\ x \in \mathscr{S}_k\}, \tag{3.63}$$

and

$$\Lambda := \cup_{k=0}^{\infty}\Lambda_k. \tag{3.64}$$

Assumption 3.5 $\Delta_k(\alpha, x) > 0$ uniformly on Λ for any optimal policy u^*.

Intuitively, this assumption almost only requires that as $k \to \infty$, for any two policies u and u', the differences of the transition probabilities $P_k^{\alpha_k(x)}(y|x) - P_k^{\alpha'_k(x)}(y|x)$, $x \in \mathscr{S}_k$, $y \in \mathscr{S}_{k+1}$, do not approach zero.

First, we prove a lemma.

Lemma 3.12 *With Assumptions 3.3 and 3.5, all the bias optimal policies are in $\mathscr{C}_0$; i.e., (3.51) holds for all $k = 0, 1, \ldots$.*

Proof The proof is based on the bias difference formula (3.60). Given any policy $u \in \mathscr{D}_0 - \mathscr{C}_0$, with $u^* \in \mathscr{C}_0$ satisfying (3.51), we prove that there is another policy in $\mathscr{C}_0$ which has a larger bias than u. Denote $u = \{\alpha_0(x), \alpha_1(x), \ldots\}$, with $\alpha_k(x)$ being the action taken at time k when the state is x. By Lemma 3.8 and (3.46),

$$\sum_y [P_k^{\alpha_k(x)}(y|x) g_{k+1}^{u^*}(y)] + f_k^{\alpha_k(x)}(x) = \sum_y [P_k^{\alpha_k^*(x)}(y|x) g_{k+1}^{u^*}(y)] + f_k^{\alpha_k^*(x)}(x),$$

$\alpha_k^* := \alpha_k^{u^*}$, holds on every k except for a non-frequently visited sequences $k_0, k_1, \ldots$ and states $x_{k_0}, x_{k_1}, \ldots$ on which

$$\begin{aligned} &\sum_y [P_{k_l}^{\alpha_{k_l}(x_{k_l})}(y|x_{k_l}) g_{k_l+1}^{u^*}(y)] + f_{k_l}^{\alpha_{k_l}(x_{k_l})}(x_{k_l}) \\ &< \sum_y [P_{k_l}^{\alpha_{k_l}^*(x_{k_l})}(y|x_{k_l}) g_{k_l+1}^{u^*}(y)] + f_{k_l}^{\alpha_{k_l}^*(x_{k_l})}(x_{k_l}). \end{aligned} \tag{3.65}$$

$l = 0, 1, \ldots$, and by Assumption 3.5 and Lemma 3.8, (3.45) must hold.

Now, we construct another policy $u' = \{\alpha'_0, \alpha'_1, \ldots, \}$ by setting

$$\alpha'_k(x) := \begin{cases} \alpha_k(x), & if\ k \neq k_l\ or\ x \neq x_{k_l},\ l = 0, 1, \ldots, \\ \alpha_k^*(x), & if\ k = k_l\ and\ x = x_{k_l},\ l = 0, 1, \ldots, \end{cases} \tag{3.66}$$

Then, $u' \in \mathscr{C}_0$. By Lemma 3.9, $\hat{g}^{u^*} - \hat{g}^{u'} = b_k e$, and u^* can be replaced by u' in (3.46) to determine $\mathscr{C}_0$. By (3.49) and (3.65), we have

$$\begin{aligned} &\sum_y [P_{k_l}^{\alpha_{k_l}(x_{k_l})}(y|x_{k_l}) g_{k_l+1}^{u'}(y)] + f_{k_l}^{\alpha_{k_l}(x_{k_l})}(x_{k_l}) \\ &< \sum_y [P_{k_l}^{\alpha'_{k_l}(x_{k_l})}(y|x_{k_l}) g_{k_l+1}^{u'}(y)] + f_{k_l}^{\alpha'_{k_l}(x_{k_l})}(x_{k_l}). \end{aligned} \tag{3.67}$$

Next, setting $u := u'$ and $u' = u$ in the difference formula (3.60), we get

$$
\begin{aligned}
&\hat{g}_0(x) - \hat{g}'_0(x) \\
&= \lim_{K\to\infty} \frac{1}{K} \sum_{k=0}^{K-1} E\{[A_k w'_k(X_k) - A'_k w'_k(X_k)] \Big| X_0 = x\} \\
&+ \lim_{K\to\infty} \frac{1}{K} \sum_{k=1}^{K-1} (K-k-1) \Big[E\Big\{ [(P_{k-1}\hat{g}'_k + f_{k-1})(X_{k-1}) \\
&- (P'_{k-1}\hat{g}'_k + f'_{k-1})(X_{k-1})] \Big| X_0 = x \Big\}\Big],
\end{aligned}
$$

in which $(P'_{k-1}\hat{g}'_k + f'_{k-1})(y) = (P_{k-1}\hat{g}'_k + f_{k-1})(y)$ for all $y \in \mathscr{S}_k, k \neq k_l$, or $y \neq x_{k_l}, l = 0, 1, \ldots$. By construction (3.66), A'_k and $A_k, k = 0, 1, \ldots$, differ only at a non-frequently visited sequence satisfying (3.45), which does not change the value of the long-run average. So the first term on the right-hand side of the above equation is zero. Next, by the construction (3.66) and (3.67), for any $k = k_l$, there is an $x_0 \in \mathscr{S}_0$ such that $\mathscr{P}[X_{k_l} = x_{k_l} | X_0 = x_0] > 0$ (cf. (3.43)). So, for this initial state x_0, the second term of the right-hand side of the above equation is negative. Thus $\hat{g}_0(x_0) < \hat{g}'_0(x_0)$; and u is not bias optimal. □

Theorem 3.5 (Bias optimality conditions) *Under Assumptions 1.1, 1.2, 3.3, 3.4, and 3.5, a policy in the space of average-reward optimal policies $u^* \in \mathscr{D}_0$ is bias optimal, if and only if*

(a) The biases $\hat{g}_k^{u^*}$, $k = 0, 1, \ldots, u^* \in \mathscr{D}_0$, satisfy

$$
(A_k^{u^*} \hat{g}_k^{u^*} + f_k^{u^*})(y) = \max_{\alpha \in \mathscr{A}_k(y)} \{(A_k^{\alpha} \hat{g}_k^{u^*} + f_k^{\alpha})(y)\},
$$

for all $y \in \mathscr{S}_k$; and

(b) The bias potentials $w_k^{u^*}$, $k = 0, 1, \ldots, u^* \in \mathscr{D}_0$, satisfy

$$
(A_k^{u^*} w_k^{u^*})(y) = \max_{\alpha \in \mathscr{A}_{0,k}(y)} \{(A_k^{\alpha} w_k^{u^*})(y)\}, \quad \forall y \in \mathscr{S}_k, \tag{3.68}
$$

on every "frequently visited" subsequence of $k = 0, 1, \ldots$; or more precisely, if there exist a subsequence of $k = 0, 1, \ldots$, denoted by $k_0, k_1, \ldots, k_l, \ldots$, a sequence of state $x_{k_0}, x_{k_1}, \ldots, x_{k_l} \in \mathscr{S}_{k_l}, l = 0, 1, \ldots$, and a sequence of actions $\alpha_{k_0}, \alpha_{k_1}, \ldots, \alpha_{k_l} \in \mathscr{A}_{0,k_l}(x_{k_l})$, $l = 0, 1, \ldots$, such that (3.68) does not hold on k_l, $l = 0, 1, \ldots$, i.e., $(A_{k_l}^{u^*} w_{k_l}^{u^*})(x_{k_l}) < (A_{k_l}^{\alpha_{k_l}} w_{k_l}^{u^*})(x_{k_l})$, then it holds that

$$
\lim_{n\to\infty} \frac{1}{k_n} \sum_{l=1}^{n} \Big\{ (A_{k_l}^{u^*} w_{k_l}^{u^*})(x_{k_l}) - (A_{k_l}^{\alpha_{k_l}} w_{k_l}^{u^*})(x_{k_l}) \Big\} = 0.
$$

Proof With Lemma 3.12 and Assumption 3.5, the bias optimal policy is in $\mathscr{C}_0$, and we may use the difference formula (3.61). It is easy to verify that conditions (a) and (b) are sufficient. With Assumption 3.5 and by Lemma 3.12, condition (a) is necessary. To prove that condition (b) is also necessary, follow the same construction procedure as in Theorem 3.2. □

In summary, to find a bias optimal policy, we first find an average-reward optimal policy $u^* = (\mathbb{P}^*, f^*)$ satisfying (3.51), and then use (3.44) to determine the space $\mathscr{C}_0 \subset \mathscr{D}_0$ in (3.48); then, a bias optimal policy can be obtained by the optimality equation (3.68) and the Poisson equation (3.54), in a similar way as for the average reward optimal policies. Analytical, numerical, or simulation methods can be developed, and these are further research topics. The results extend those for THMCs (Altman 1999; Bertsekas 2007; Cao 2007; Feinberg and Shwartz 2002; Hernández-Lerma and Lasserre 1996, 1999; Jasso-Fuentes and Hernández-Lerma 2009b; Lewis and Puterman 2001, 2002; Puterman 1994).

3.5 Conclusions

In this chapter, we have shown that the notion of confluencity serves as the base for the optimization of Markov chains for infinite horizon problems. With confluencity, relative potentials and performance potentials can be defined; the former measures the change in total reward due to a shift in the states, and the latter represents the contribution of a state to the average reward. The performance difference formula was derived, from which the necessary and sufficient optimality equations were obtained simply by comparing the average rewards of any two policies.

In the time-nonhomogeneous case, the average performance criterion is under-selective, in the sense that (1) the optimality conditions may not need to hold on a sequence of "non-frequently visited" time periods, including any finite periods; and (2) the total reward in any finite period of an optimal policy may not be the best. The first issue is naturally demonstrated by the comparison of the performance measures on the entire infinite horizon of $[0, \infty)$, and the issue has been successfully addressed in the optimality conditions in Theorem 3.2. To overcome the second issue of the under-selectivity of the average-reward criterion, we have studied the bias optimality for transient performance; we have shown that for the bias to be optimal, one additional condition based on the bias potentials is required. Higher level features of under-selectivity are reflected by the Nth biases, $N = 1, 2, \ldots$, which will be studied in Chap. 5.

3.6 Appendix

A. Proof of Lemma 3.3:

First, we define $\tilde{\tau}_k(x, y)$ as the total time required to reach any point at which $\boldsymbol{X}$ and $\boldsymbol{X}'$ take the same value (not necessary the first point as defined in $\tau_k(x, y)$); i.e., set

$$\tilde{\tau}_k(x, y) \in \{all\ \tau - k : \tau \geq k\ with\ X_\tau = X'_\tau,\\ X_k = x, X'_k = y\}, \qquad k = 0, 1, \ldots, \tag{3.69}$$

This $\tilde{\tau}_k(x, y)$ is not unique. Because $\boldsymbol{X}$ and $\boldsymbol{X}'$ are statistically identical between $k + \tau_k(x, y)$ and any $k + \tilde{\tau}_k(x, y)$, by symmetry, we have that for $k = 0, 1, \ldots$,

$$E\Big\{ \sum_{l=k+\tau_k(x,y)}^{k+\tilde{\tau}_k(x,y)} [f_l(X'_l) - f_l(X_l)] \Big| X'_k = y, X_k = x \Big\} = 0.$$

Thus, for any $\tilde{\tau}_k(x, y) > \tau_k(x, y)$, we have (cf. (3.6))

$$\gamma_k(x, y) = E\Big\{ \sum_{l=k}^{k+\tilde{\tau}_k(x,y)} [f_l(X'_l) - f_l(X_l)] \Big| X'_k = y, X_k = x \Big\}. \tag{3.70}$$

Now, we consider three independent Markov chains $\boldsymbol{X}^x$, $\boldsymbol{X}^y$, and $\boldsymbol{X}^z$, with the same $\mathbb{P}$ and $\boldsymbol{f}$, starting at time k with three different states x, y, and z, in $\mathscr{S}_k$, respectively. If $\tau_k(x, z) = \tau_k(x, y)$, then $\boldsymbol{X}^y_{\tau_k(x,y)} = \boldsymbol{X}^x_{\tau_k(x,y)} = \boldsymbol{X}^z_{\tau_k(x,z)}$, and thus $\tau_k(y, z) \leq \tau_k(x, y)$, and $\tau_k(x, y)$ is one of the $\tilde{\tau}_k(y, z)$'s defined in (3.69). Thus, we have

$$\begin{aligned} &E\Big\{ \sum_{l=k}^{k+\tau_k(x,y)} [f_l(X^y_l) - f_l(X^x_l)] \Big| \tau_k(x, z) = \tau_k(x, y), X^y_k = y, X^x_k = x \Big\} \\ = &E\Big\{ \sum_{l=k}^{k+\tilde{\tau}_k(y,z)} [f_l(X^y_l) - f_l(X^z_l)] \Big| \tau_k(x, z) = \tau_k(x, y), X^y_k = y, X^z_k = z \Big\} \\ &+ E\Big\{ \sum_{l=k}^{k+\tau_k(x,z)} [f_l(X^z_l) - f_l(X^x_l)] \Big| \tau_k(x, z) = \tau_k(x, y), X^z_k = z, X^x_k = x \Big\}. \end{aligned} \tag{3.71}$$

By (3.70), this is (3.12) in Lemma 3.3 (with $\tau_k(x, z) = \tau_k(x, y)$).

Now, we assume $\tau_k(x, z) < \tau_k(x, y)$. Set

$$k + \tilde{\tau}_k(z, y) := \min\{\tau > k + \tau_k(x, z), X^z_\tau = X^y_\tau\}.$$

By the Markov property, $\boldsymbol{X}^x$ and $\boldsymbol{X}^z$ are statistically identical after $k + \tau_k(x, z)$; so

$$E\Big\{\sum_{l=k+\tau_k(x,z)}^{k+\tau_k(x,y)} [f_l(X_l^x) - f_l(X_l^y)]\Big| X_{k+\tau_k(x,z)}^x = X_{k+\tau_k(x,z)}^z, \tau_k(x,z) < \tau_k(x,y)\Big\}$$

$$= E\Big\{\sum_{l=k+\tau_k(x,z)}^{k+\tilde{\tau}_k(z,y)} [f_l(X_l^z) - f_l(X_l^y)]\Big| X_{k+\tau_k(x,z)}^x = X_{k+\tau_k(x,z)}^z, \tau_k(x,z) < \tau_k(x,y)\Big\}.$$

From this and other similar equations due to the Markov property, we can verify that

$$\begin{aligned} &E\Big\{\sum_{l=k}^{k+\tau_k(x,y)} [f_l(X_l^y) - f_l(X_l^x)]\Big| \tau_k(x,z) < \tau_k(x,y), X_k^y = y, X_k^x = x\Big\} \\ &= E\Big\{\sum_{l=k}^{k+\tilde{\tau}_k(y,z)} [f_l(X_l^y) - f_l(X_l^z)]\Big| \tau_k(x,z) < \tau_k(x,y), X_k^y = y, X_k^z = z\Big\} \\ &+E\Big\{\sum_{l=k}^{k+\tau_k(x,z)} [f_l(X_l^z) - f_l(X_l^x)]\Big| \tau_k(x,z) < \tau_k(x,y), X_k^z = z, X_k^x = x\Big\} \end{aligned} \tag{3.72}$$

By (3.11), this is (3.12) in Lemma 3.3 under the condition $\tau_k(x,z) < \tau_k(x,y)$. Finally, if $\tau_k(x,z) > \tau_k(x,y)$, then (3.12) can be proved by exchanging y and z in (3.72).

The lemma follows directly from (3.71)–(3.72). □

B. Proof of Lemma 3.4:

Let g_k and g'_k, $k = 0, 1, \ldots$, be the solutions to the Poisson equation (3.14) with two different sets of constants c_k and c'_k, $k = 0, 1, \ldots$, respectively; i.e.,

$$g_k - P_k g_{k+1} - f_k = -c_k e, \quad and \quad g'_k - P_k g'_{k+1} - f_k = -c'_k e.$$

Setting $h_k = g'_k - g_k$ and $d_k = c'_k - c_k, k = 0, 1, \ldots$, we have $h_k - P_k h_{k+1} = -d_k e$. Now, define $h'_0 = h_0, h'_{k+1} = h_{k+1} - (\sum_{l=0}^{k} d_l)e, k = 0, 1, \ldots$. Then, we have $h'_k - P_k h'_{k+1} = 0, k = 0, 1, \ldots$. Therefore, $h'_0 = (\prod_{l=0}^{K} P_l)h'_{K+1}$. This holds for all K, and thus $\lim_{K\to\infty} E\{h'_K(X_K)|X_0 = x\} = h'_0(x)$ for all $x \in \mathscr{S}_0$. However, by Lemma 2.1, as $K \to \infty$, $E\{h'_K(X_K)|X_0 = x\}$ does not depend on x; so we must have $h'_0 = ce$ with $c = \lim_{K\to\infty} E\{h'_K(X_K)|X_0 = x\}$. Starting from any k, we can prove $h'_k = ce$ for any k, and thus

$$g'_k = g_k + (c + \sum_{l=0}^{k-1} d_l)e = g_k + b_k e, \quad k = 0, 1, \ldots,$$

with $\sum_{l=0}^{-1} d_l := 0$, and

$$\begin{aligned}
b_k &= c + \sum_{l=0}^{k-1} d_l \\
&= \lim_{K\to\infty} E[g'_K(X_K) - g_K(X_K)|X_0 = x] \\
&\quad - \sum_{l=k}^{\infty}(c'_l - c_l).
\end{aligned}$$

The lemma is proved. □

C. Proof of Lemma 3.5:

By (3.19), we only need to prove (see Footnote 3 on Page 34)

$$\lim_{K\to\infty} E[g_K(X_K)|X_k = y] = 0, \ \ \forall y \in \mathscr{S}_k. \tag{3.73}$$

The proof is similar to Lemma 2.1. With (2.3), we have the bound (3.20), and for any $x, x' \in \mathscr{S}_k$,

$$\begin{aligned}
&\Big|E[g_K(X_K)|X_k = x] - E[g_K(X_K)|X_k = x']\Big| \\
&= \Big|\sum_{z\in\mathscr{S}_K} g_K(z)\mathscr{P}(X_K = z, \tau_k(x,x') > K-k|X_k = x) \\
&\quad - \sum_{z\in\mathscr{S}_K} g_K(z)\mathscr{P}(X_K = z, \tau_k(x,x') > K-k|X_k = x')\Big| \\
&\quad\quad \times \mathscr{P}(\tau_k(x,x') > K-k) \\
&\le 2M_1 M \mathscr{P}(\tau_k(x,x') > K-k). \qquad (S_K < M_1)
\end{aligned}$$

Therefore, (3.73) and (3.21) hold for $k = 0$, since $E[g_K(X_K)|X_0 = x_0] = 0$ for all K, as shown in (3.16).

Next, for any $k > 0$, we have

$$\begin{aligned}
&\Big|E[g_{k+K}(X_{k+K})|X_0 = x] - E[g_{k+K}(X_{k+K})|X_k = x']\Big| \\
&= \Big|\sum_{z\in\mathscr{S}_k} E[g_{k+K}(X_{k+K})|X_k = z]\mathscr{P}(X_k = z|X_0 = x) \\
&\quad\quad - E[g_{k+K}(X_{k+K})|X_k = x']\Big| \\
&\le \sum_{z\in\mathscr{S}_k} \Big\{\Big|E[g_{k+K}(X_{k+K})|X_k = z] \\
&\quad - E[g_{k+K}(X_{k+K})|X_k = x']\Big|\mathscr{P}(X_k = z|X_0 = x)\Big\}.
\end{aligned}$$

Thus, with the strong confluencity, (3.73), and hence (3.21), holds for all $k > 0$. □

D. Proof of Lemma 3.11:

By the Poisson equation (3.42), we have $\hat{g}_k - P_k\hat{g}_{k+1} - f_k = \hat{g}'_k - P'_k\hat{g}'_{k+1} - f'_k,\ k = 0, 1, \ldots$. Recursively, we can prove

$$\hat{g}'_0 - \hat{g}_0 = (\prod_{l=0}^{k-1} P'_l)[\hat{g}'_k - \hat{g}_k]$$
$$+\sum_{l=1}^{k}\Big\{(\prod_{i=0}^{l-2} P'_i)[(P'_{l-1}\hat{g}_l + f'_{l-1}) - (P_{l-1}\hat{g}_l + f_{l-1})]\Big\},$$

with the convention $\prod_{i=0}^{-1} P'_i = I$. Componentwise, for any $x \in \mathscr{S}_0$, it holds that for any $k \geq 0$,

$$\hat{g}'_0(x) - \hat{g}_0(x) = E'[\hat{g}'_k(X'_k) - \hat{g}_k(X'_k)|X'_0 = x]$$
$$+\sum_{l=1}^{k} E'\Big\{[(P'_{l-1}\hat{g}_l + f'_{l-1})(X'_{l-1})$$
$$-(P_{l-1}\hat{g}_l + f_{l-1})(X'_{l-1})]\Big|X'_0 = x\Big\}.$$

Averaging it over 0 to $K - 1$ and taking the limit, we have

$$\hat{g}'_0(x) - \hat{g}_0(x)$$
$$= \lim_{K\to\infty}\frac{1}{K}\sum_{k=0}^{K-1} E'[\hat{g}'_k(X'_k) - \hat{g}_k(X'_k)|X'_0 = x]$$
$$+\lim_{K\to\infty}\frac{1}{K}\sum_{k=1}^{K-1}\sum_{l=1}^{k} E'\Big\{[(P'_{l-1}\hat{g}_l + f'_{l-1})(X'_{l-1})$$
$$-(P_{l-1}\hat{g}_l + f_{l-1})(X'_{l-1})]\Big|X'_0 = x\Big\}. \tag{3.74}$$

Next, by (3.55) we have

$$\lim_{K\to\infty}\frac{1}{K}\sum_{k=0}^{K-1} E'[\hat{g}'_k(X'_k) - \hat{g}_k(X'_k)|X'_0 = x] \tag{3.75}$$
$$= \lim_{K\to\infty}\frac{1}{K}\sum_{k=0}^{K-1} E'[A'_k w_k(X'_k) - A_k w_k(X'_k)|X'_0 = x]$$
$$+\lim_{K\to\infty}\frac{1}{K}\sum_{k=0}^{K-1} E'[A'_k w'_k(X'_k) - A'_k w_k(X'_k)|X'_0 = x].$$

Applying Dynkin's formula, we get

$$\sum_{k=0}^{K-1} E'[A'_k w'_k(X'_k) - A'_k w_k(X'_k)|X'_0 = x]$$
$$= E'[w'_K(X'_K) - w_K(X'_K)|X'_0 = x] - [w'_0(X'_0) - w_0(X'_0)].$$

If Assumption 3.4 and the strong confluencity hold, from Lemma 3.10, (3.56), and (3.58), we have

$$\lim_{K\to\infty} \frac{1}{K} \sum_{k=0}^{K-1} E'[A'_k w'_k(X'_k) - A'_k w_k(X'_k)|X'_0 = x] = 0. \tag{3.76}$$

Finally, from (3.74), (3.75), and (3.76), we have

$$\begin{aligned} &\hat{g}'_0(x) - \hat{g}_0(x) \\ &= \lim_{K\to\infty} \frac{1}{K} \sum_{k=0}^{K-1} E'\Big\{[A'_k w_k(X'_k) - A_k w_k(X'_k)]\Big| X'_0 = x\Big\} \\ &\quad + \lim_{K\to\infty} \frac{1}{K} \sum_{k=1}^{K-1} \sum_{l=1}^{k} E'\Big\{[(P'_{l-1}\hat{g}_l + f'_{l-1})(X'_{l-1}) \\ &\quad - (P_{l-1}\hat{g}_l + f_{l-1})(X'_{l-1})]\Big| X'_0 = x\Big\}. \end{aligned}$$

Exchanging the order of the two summations over k and l in the last equation, we get the *bias difference formula* (3.60) (cf. Eq. (4.54) in Cao 2007). □

Chapter 4
Optimization of Average Rewards: Multi-Chains

In this chapter, we study the optimization of the long-run average of multi-class TNHMCs. For simplicity, we assume that there is no state joining the Markov chain at $k > 0$, and therefore (3.34) holds, i.e., $\mathscr{S}_{k-1,out} = \mathscr{S}_k$, $k = 1, 2, \ldots$. In this case, the number of confluent classes is fixed, i.e., $d_k = d$, $k = 0, 1, \ldots$.

We show that with confluencity, state classification, and relative optimization, we can obtain the necessary and sufficient conditions for optimal policies of the average reward of TNHMCs consisting of multiple confluent classes (multi-chai). Just like in the uni-chain TNHMC case discussed in Chap. 3, we apply the relative optimization approach; and the optimality conditions do not need to hold in any finite period, or "non-frequently visited" time sequence.

In the main part of this chapter, we assume that the limit of the average exists. In general, the performance should be defined as the "lim inf" of the average, as in (1.6). However, because of the non-linear property of "lim inf", it is not well defined for branching states, unless the TNHMC is "asynchronous" among different confluent classes. This property will be briefly discussed in Sect. 4.4.

4.1 Performance Potentials of Multi-class TNHMCs

We consider a multi-class TNHMC with d multiple confluent classes, with

$$\mathscr{R}_k := \cup_{r=1}^{d} \mathscr{R}_{k,r} \; and \; \mathscr{R}_{\cdot r} := \cup_{k=0}^{\infty} \mathscr{R}_{k,r}$$

being the set of all confluent states at k and the set of all rth confluent states at all times k, $k = 0, 1, \ldots$, respectively (see (2.10) and (2.11)), and

X.-R. Cao, *Foundations of Average-Cost Nonhomogeneous Controlled Markov Chains*, SpringerBriefs in Control, Automation and Robotics,
https://doi.org/10.1007/978-3-030-56678-4_4

$$\mathscr{R} := \cup_{k=0}^{\infty} \mathscr{R}_k = \cup_{r=1}^{d} \mathscr{R}_{\cdot r}$$

being the set of all confluent states.

The set of all branching states at time k is denoted by $\mathscr{T}_k$, and the set of branching states at all times is (see, (2.12))

$$\mathscr{T} = \cup_{k=0}^{\infty} \mathscr{T}_k.$$

Similar to (1.6), we define the performance measure as (assume it exists!)

$$\eta_k(x) = \lim_{K\to\infty} \frac{1}{K} E\Big\{ \sum_{l=k}^{k+K-1} f_l(X_l) \Big| X_k = x \Big\}, \quad x \in \mathscr{S}_k, \tag{4.1}$$

for $k = 0, 1, \ldots$. Set the vector $\eta_k = (\eta_k(1), \ldots, \eta_k(S_k))^T$, with $\eta_0(x) = \eta(x)$. Then, we have (see (3.2))

$$\eta_k = P_k \eta_{k+1}. \tag{4.2}$$

Recursively applying (4.2) yields

$$\eta_0 = (P_0 \cdot P_1 \ldots P_{k-1}) \eta_k.$$

Componentwise, this is

$$\eta_0(x) = \sum_{y \in \mathscr{S}_k} \eta_k(y) \mathscr{P}(X_k = y | X_0 = x), \tag{4.3}$$

and from which, we have

$$\eta_0(x) = E[\eta_K(X_K)|X_0 = x], \quad for\ any\ K > 0.$$

In general, we have

$$\eta_k(x) = E[\eta_K(X_K)|X_k = x], \quad for\ any\ K > k \geq 0. \tag{4.4}$$

Lemma 4.1 *Under Assumption 1.1, the average rewards defined in (4.1) are the same for all the states in the same confluent class; i.e.,* $\eta_k(x) \equiv \eta_{\cdot r}$, *for all* $x \in \mathscr{R}_{k,r}$, $r = 1, 2, \ldots, d$, $k = 0, 1, \ldots$.

Proof This lemma is a special case of Lemma 3.2. There is, however, another more straightforward proof when the "$\lim_{k\to\infty}$" in the long-run average exists.

From (4.3), for the long-run average (4.1), we have

$$
\begin{aligned}
&|\eta_0(x) - \eta_0(x')| \\
&= \Big| \sum_{y \in \mathscr{S}_k} \eta_k(y) \mathscr{P}(X_k = y | X_0 = x) - \sum_{y \in \mathscr{S}_k} \eta_k(y) \mathscr{P}(X_k = y | X_0 = x') \Big| \\
&\le \sum_{y \in \mathscr{S}_k} |\eta_k(y)| \times \Big| \mathscr{P}(X_k = y | X_0 = x) - \mathscr{P}(X_k = y | X_0 = x') \Big|.
\end{aligned}
$$

With (1.10), $\eta_k(x)$ is bounded for all k; and therefore, by Lemma 2.1 (weak ergodicity in confluent class r),

$$
\eta_0(x) = \eta_0(x') =: \eta_{\cdot r}, \quad x, x' \in \mathscr{R}_{0,r}, \ r = 1, 2, \ldots, d.
$$

Similarly, we can prove $\eta_k(x) = \eta_{\cdot r}$ for all $x \in \mathscr{R}_{k,r}$, $r = 1, 2, \ldots, d$, $k = 0, 1 \ldots$.
□

Equation (4.1) is equivalent to

$$
\lim_{K \to \infty} \frac{1}{K} \sum_{l=k}^{k+K-1} E\{[f_l(X_l) - \eta_{\cdot r}] | X_k = x\} = 0,
$$

for any $x \in \mathscr{R}_{k,r}$. The sum $\lim_{K \to \infty} \sum_{l=k}^{k+K-1} E\{[f_l(X_l) - \eta_{\cdot r}] | X_k = x\}$ is called a *bias*. The following assumption makes a requirement on the convergence rate of (4.1).

Assumption 4.6 (a) Every TNHMC is strongly connected (see Definition 2.5).
(b) The biases for all confluent classes in any finite period are bounded, i.e.,

$$
\Big| \sum_{l=k}^{k+K-1} E\{[f_l(X_l) - \eta_{\cdot r}] | X_k = x\} \Big| < M_4,
$$

for all $x \in \mathscr{R}_{k,r}$, $r = 1, \ldots, d$, $k = 1, 2, \ldots$, and $K = 0, 1, \ldots$.

Now, we apply the results of Chap. 3 to each confluent class. Consider two independent sample paths, $\{X_l, l \ge k\}$ and $\{X'_l, l \ge k\}$, with the same transition law $\mathbb{P} = \{P_l, l = k, k+1, \ldots\}$ and starting from two different initial states at time k in the same confluent class $X_k = x$, $X'_k = y$, $x, y \in \mathscr{R}_{k,r}$, $r = 1, 2, \ldots, d$. Assume that the strong connectivity holds for all the confluent classes (Definition 2.3).

The *relative performance potential* for any two confluent states in the same confluent class $x, y \in \mathscr{R}_{k,r}$, $r = 1, 2, \ldots, d$, is defined by (see (3.6) in Sect. 2.1):

$$
\gamma_{k,r}(x, y) := E\Big\{ \sum_{l=k}^{k+\tau_k(x,y)} [f_l(X'_l) - f_l(X_l)] \Big| X'_k = y, X_k = x \Big\}, \quad k = 0, 1, \ldots.
$$

By the strong connectivity and Assumption 1.1(b), $|\gamma_{k,r}(x, y)| < 2M_2M_3 < \infty$ holds.

As in (3.13), there is a *performance potential function* $g_{k,r}(x)$, $x \in \mathscr{R}_{k,r}$, such that

$$\gamma_{k,r}(x, y) = g_{k,r}(y) - g_{k,r}(x), \quad k = 0, 1, \ldots.$$

Let $g_{k,r} = (g_{k,r}(1), \ldots, g_{k,r}(|\mathscr{R}_{k,r}|))^T$ be the vector of potentials for the rth confluent class at time k, As in (3.14), there must be a constant $c_{k,r}$ such that

$$g_{k,r} - P_{k,r} g_{k+1,r} - f_{k,r} = -c_{k,r} e, \tag{4.5}$$

$r = 1, 2, \ldots, d,\ k = 0, 1, \ldots,$ and $P_{k,r}$ is the state-transition probability matrix among the states in $\mathscr{R}_{\cdot r}$ at time k. This is the *Poisson equation*; it also has the form

$$A_{k,r} g_{k,r}(x) + f_{k,r}(x) = c_{k,r}, \ r = 1, 2, \ldots, d, \ k = 1, 2, \ldots,$$

where $A_{k,r}$ is the infinitesimal generator defined on any sequence of functions $h_{k,r}(x)$, $x \in \mathscr{R}_{k,r}$, $r = 1, 2, \ldots, d$, $k = 0, 1, \ldots,$ as follows

$$A_{k,r} h_{k,r}(x) = \sum_{y \in \mathscr{R}_{k+1,r}} P_{k,r}(y|x) h_{k+1,r}(y) - h_{k,r}(x).$$

As discussed for (3.14), $c_{k,r}$ in (4.5) can be any real number. In addition, for any fixed $c_{k,r}$, the biases $g_{k,r}$ and $g_{k+1,r}$ (for states in the same $\mathscr{R}_r$) are unique only up to an additional constant. It is shown in Lemma 3.4 that for any k and r the solutions to (4.5) corresponding to two different $c_{k,r}$'s differ only by a constant vector, and hence they lead to the same $\gamma_{k,r}(x, y)$ for all $x, y \in \mathscr{R}_{k,r}$.

The sample paths staring from states in different confluent classes do not meet; therefore, $c_{k,r}$ may be chosen separately for different classes. We choose

$$c_{k,r} = \eta_{\cdot r}, \quad k = 0, 1, \ldots.$$

Because $\eta_k(x) = \eta_{\cdot r}$ for any $x \in \mathscr{R}_{k,r}$, $r = 1, 2, \ldots, d$, we can write

$$A_{k,r} g_{k,r}(x) + f_{k,r}(x) = \eta_k(x), \quad x \in \mathscr{R}_{k,r}. \tag{4.6}$$

For branching states, relative potentials cannot be defined in the same way as for confluent states, because (3.6) is not well defined (in other words, $\tau_k(x, y)$ is not finite if x is branching). However, from a branching state the Markov chain eventually reaches one of the confluent classes, so we can define performance potentials for branching states by using those of the confluent states. Specifically, for any branching state $x \in \mathscr{T}_k$, following (4.6), we define

$$g_k(x) - \sum_{y \in \mathscr{S}_{k+1}} P_k(y|x) g_{k+1}(y) - f_k(x) = -\eta_k(x). \tag{4.7}$$

In (4.7), $g_k(x)$ is determined by the $g_{k+1}(y)$'s, with $y \in \mathscr{S}_{k+1}$ being reachable from x. If y is confluent, then $g_{k+1}(y)$ is known via (4.6); if y is branching, $g_{k+1}(y)$ can be determined by the $g_{k+2}(z)$'s, with $z \in \mathscr{S}_{k+2}$ being reachable from y, and so on. Because starting from any $x \in \mathscr{T}_k$, eventually, only confluent states are reachable, $g_k(x)$, $x \in \mathscr{T}_k$, can be determined by the potentials of confluent states via (4.7). By the strong connectivity and because of (2.14), all $g_k(x)$, $x \in \mathscr{S}_k$, for all k, are finite.

Let $g_k := (g_k(1), g_k(2), \ldots, g_k(S_k))^T$ be the vector of potentials of all states at time k. Combining (4.6) and (4.7), we get the *Poisson equation* for all $x \in \mathscr{S}_k$:

$$g_k - P_k g_{k+1} - f_k = -\eta_k, \quad k = 0, 1, \ldots; \tag{4.8}$$

or with the generator $A_k h_k(x)$, it is

$$A_k g_k(x) + f_k(x) = \eta_k(x), \ x \in \mathscr{S}_k. \tag{4.9}$$

From (4.8), we have for any $K > k$ (Cao 2015),

$$g_k = \sum_{n=k}^{K-1} \Big[(\prod_{l=k}^{n-1} P_l)(f_n - \eta_n) \Big] + (\prod_{n=k}^{K-1} P_n) g_K,$$

with $\prod_{l=k}^{k-1} P_l = I$ as a conversion. For each component $x \in \mathscr{S}_k$, we have (cf. (3.19))

$$g_k(x) = \sum_{l=k}^{K-1} E[f_l(X_l) - \eta_l(X_l)|X_k = x] \\ + E[g_K(X_K)|X_k = x]. \tag{4.10}$$

Lemma 4.2 *With the strong connectivity and under Assumptions 1.1 and 4.6, $g_k(x)$, $x \in \mathscr{S}_k$, $k = 0, 1, \ldots$, are bounded, i.e., $|g_k(x)| < M_5$, for all $x \in \mathscr{S}_k$, $k = 0, 1, \ldots$.*

Proof By (4.10) and Assumption 4.6, for any confluent state $x \in \mathscr{R}_{0,r}$, $r = 1, \ldots, d$, we have $|E[g_K(X_K)|X_0 = x]| < M_4 + |g_0(x)| =: L$, for any $K > 0$. Thus, when K is large enough, by the weak ergodicity, $|E[g_K(X_K)|X_k = x]|$, $K \gg k$, are very close to each other for different $x \in \mathscr{R}_{k,r}$ and different k, and so $|E[g_K(X_K)|X_k = x]| < 2L$. By (4.10) again, for any $k > 0$ and $x \in \mathscr{R}_{k,r}$, we have $|g_k(x)| < 3L$.

Next, for any $x \in \mathscr{T}_k$, from (4.10), we have

$$\begin{aligned} g_k(x) &= E\Big\{ \sum_{l=k}^{\tau_{k,\mathscr{R}}(x)-1} [f_l(X_l) - \eta_l(X_l)] \Big| X_k = x \Big\} \\ &\quad + E[g_{\tau_{k,\mathscr{R}}(x_k)}(X_{\tau_{k,\mathscr{R}}(x_k)})|X_k = x_k] \\ &= \sum_{r=1}^{d} \Big\{ p_{k,r}(x_k) \Big[E\{ \sum_{l=k}^{\tau_{k,\mathscr{R}}(x_k)-1} [f_l(X_l) - \eta_l(X_l)] \end{aligned}$$

$$|X_k = x_k, X_{\tau_{k,\mathscr{R}}} \in \mathscr{R}_{\cdot,r}\}$$
$$+ E[g_{\tau_{k,\mathscr{R}}(x_k)}(X_{\tau_{k,\mathscr{R}}(x_k)})|X_k = x_k, X_{\tau_{k,\mathscr{R}}} \in \mathscr{R}_{\cdot,r}]\Big]\Big\}.$$

Therefore, for any $x \in \mathscr{T}$ we have $|g_k(x)| < d[2M_2M_3 + 3L]$. Thus, $|g_k(x)|$ are bounded for all $x \in \mathscr{S}_k$, $k = 0, 1, \ldots$. □

4.2 Optimization of Average Rewards: Multi-class

4.2.1 The Performance Difference Formula

Consider two independent Markov chains $\boldsymbol{X}$ and $\boldsymbol{X}'$ with $(\mathbb{P}, \boldsymbol{f})$ and $(\mathbb{P}', \boldsymbol{f}')$, $\mathbb{P} = \{P_0, P_1, \ldots\}$, $\boldsymbol{f} = (f_0, f_1, \ldots)$ and $\mathbb{P}' = \{P_0', P_1', \ldots\}$, $\boldsymbol{f}' = (f_0', f_1', \ldots)$, starting from the same initial state $X_0 = X_0' = x \in \mathscr{S}_0$; each of them is strongly connected. Assume that $\mathscr{S}_k' = \mathscr{S}_k$, $k = 0, 1, \ldots$, (but they may have different confluent and branching classes) and Assumption 1.1(b) holds. Let η_k, g_k, A_k, and η_k', g_k', A_k' be the quantities associated with $\boldsymbol{X}$ and $\boldsymbol{X}'$, respectively. Let $E := E^{\mathbb{P}}$ and $E' := E^{\mathbb{P}'}$ denote the expectations corresponding to the probability measures generated by $\mathbb{P}$ and $\mathbb{P}'$, respectively.

Applying the Poisson equation (4.9) for $\boldsymbol{X}$ to the states visited by $\boldsymbol{X}'$ leads to

$$A_k g_k(X_k') + f_k(X_k') = \eta_k(X_k'), \quad X_k' \in \mathscr{S}_k.$$

Taking the same operations on both sides, we get

$$\lim_{K\to\infty} \frac{1}{K} \sum_{k=0}^{K-1} E'[\eta_k(X_k')|X_0' = x]$$
$$= \lim_{K\to\infty} \frac{1}{K} E'\{\sum_{k=0}^{K-1} (A_k g_k + f_k)(X_k')|X_0' = x\}. \tag{4.11}$$

Applying Dynkin's formula (3.25) for $\boldsymbol{X}'$ with $h_k(x) = g_k(x)$ leads to

$$E'\{\sum_{k=0}^{K-1} [A_k' g_k(X_k')]|X_0' = x\} = E'[g_K(X_K')|X_0' = x] - g_0(x).$$

It follows from Lemma 4.2 that $\lim_{K\to\infty} \frac{1}{K} E'[g_K(X_K')]|X_k' = x] = 0$. Thus, we have

$$\lim_{K\to\infty} \frac{1}{K} E'\{\sum_{k=0}^{K-1} [A_k' g_k(X_k')]|X_0' = x\} = 0.$$

By definition,

$$\eta' = \lim_{K\to\infty} \frac{1}{K} E'\Big\{ \sum_{k=0}^{K-1} f'_k(X'_k) \Big| X'_0 = x \Big\}.$$

Adding the two sides of the above two equations together yields

$$\eta' = \lim_{K\to\infty} \frac{1}{K} E'\Big\{ \sum_{k=0}^{K-1} [(A'_k g_k(X'_k) + f'_k(X'_k)] \Big| X'_0 = x \Big\}.$$

Finally, from this equation and (4.11), we get the performance difference formula (cf. Eq. (4.36) in Cao (2007) for THMCs):

$$\begin{aligned}
&\eta'(x) - \eta(x) \\
&= \Big\{ \lim_{K\to\infty} \frac{1}{K} E'\{\sum_{k=0}^{K-1} [(P'_k g_{k+1})(X'_k) + f'_k(X'_k)] | X'_0 = x\} \\
&\quad + \lim_{K\to\infty} \frac{1}{K} \sum_{k=0}^{K-1} E'[\eta_k(X'_k) | X'_0 = x] \Big\} \\
&- \Big\{ \lim_{K\to\infty} \frac{1}{K} E'\{\sum_{k=0}^{K-1} [(P_k g_{k+1})(X'_k) + f_k(X'_k)] | X'_0 = x\} \\
&\quad + \eta(x) \Big\}.
\end{aligned} \tag{4.12}$$

If $\boldsymbol{X}$ is a uni-chain, then $\eta_k(X'_k) \equiv \eta(x)$ for all X'_k, the second terms in the two curly braces cancel out, and (4.12) becomes the same as (3.28) in Chap. 3.

4.2.2 Optimality Condition: Multi-Class

In performance optimization, the performance measure (4.1) depends on policy u:

$$\eta^u_k(x) = \lim_{K\to\infty} \frac{1}{K} E^{\mathbb{P}}\Big\{ \sum_{l=k}^{K+k-1} f^u_l(X^u_l) \Big| X^u_k = x \Big\}; \tag{4.13}$$

and we assume that the state spaces $\mathscr{S}_k$, $k = 0, 1, \ldots$, are the same for all policies.

Definition 4.1 A policy $u = (\mathbb{P}, \not{f})$ is said to be *admissible*, if Assumptions 1.1, 1.2, and 4.6 hold.

Strong connectivity is needed for $\gamma_k(x, y)$ to be bounded and for the proof of Lemma 4.2. In fact, it requires some uniformity in the behavior of the Markov chain at different times $k = 0, 1, \ldots$.

Let $\mathscr{D}$ denote the space of all admissible policies. The goal of performance optimization is to find a policy in $\mathscr{D}$ that maximizes the rewards (if it can be achieved); i.e., to identify an optimal policy denoted by u^* such that

$$u^* = \arg\left\{\max_{u\in\mathscr{D}}[\eta_k^u(x)]\right\},$$

with $\eta_k^u(x)$ defined in (4.13), for all $x \in \mathscr{S}_k$, $k = 0, 1, \ldots$. We will see that under some conditions, such an optimal policy exists.

Now we have the following *Comparison Lemma* for multi-class TNHMCs.

Lemma 4.3 *Consider any two admissible policies $u = (\mathbb{P}, \mathbb{f})$ and $u' = (\mathbb{P}', \mathbb{f}')$, with corresponding performance measures and potentials denoted by $\eta \equiv \eta_0, \eta_1, \ldots$, and $g_0, g_1, \ldots$, and $\eta' \equiv \eta'_0, \eta'_1, \ldots$, and $g'_0, g'_1, \ldots$, respectively. If*

(a) $P'_k\eta_{k+1} \le \eta_k$, for all $k = 0, 1, \ldots$; and there is an $\varepsilon > 0$ such that if

$$\eta_k(x) - P'_k\eta_{k+1}(x) > 0, \quad x \in \mathscr{S}_k, \ k = 0, 1, \ldots,$$

then

$$\eta_k(x) - P'_k\eta_{k+1}(x) > \varepsilon;$$

and

(b) If $(P'_k\eta_{k+1})(x) = \eta_k(x)$, then

$$(P'_k g_{k+1})(x) + f'_k(x) \le (P_k g_{k+1})(x) + f_k(x),$$

for any $x \in \mathscr{S}_k$, $k = 0, 1, \ldots$;

then $\eta'_k \le \eta_k$, for all $k = 0, 1, \ldots$.

Proof First, we have $E'[\eta_k(X'_k)|X'_0 = x] = P'_0(\cdot|x)(\prod_{l=1}^{k-1} P'_l)\eta_k$. From condition (a), we have $P'_{k-1}\eta_k \le \eta_{k-1}$, and thus $P'_{k-2}(P'_{k-1}\eta_k) \le P'_{k-2}\eta_{k-1} \le \eta_{k-2}$. Continuing this process yields

$$E'[\eta_k(X'_k)|X'_0 = x] \le \eta_0(x) = \eta(x), \quad k = 0, 1, \ldots.$$

Therefore,

$$\frac{1}{K}\sum_{k=0}^{K-1} E'[\eta_k(X'_k)|X'_0 = x] \le \eta(x);$$

and in the performance difference formula (4.12), we have

$$\lim_{K\to\infty}\frac{1}{K}\sum_{k=0}^{K-1} E'[\eta_k(X'_k)|X'_0 = x] \le \eta(x). \tag{4.14}$$

Next, applying Dynkin's formula, we have

$$\lim_{K\to\infty}\frac{1}{K}\sum_{k=0}^{K-1}E'\{[(P'_k\eta_{k+1})(X'_k)-\eta_k(X'_k)]|X'_0=x\}=0.$$

Let $I(x<0)=1$ if $x<0$ and $I(x=0)=1$ if $x=0$. The above equation implies

$$\lim_{K\to\infty}\frac{1}{K}\sum_{k=0}^{K-1}E'\{[(P'_k\eta_{k+1})(X'_k)-\eta_k(X'_k)]$$
$$I[(P'_k\eta_{k+1})(X'_k)-\eta_k(X'_k)]<0|X'_0=x\}=0. \tag{4.15}$$

Now, we group the terms together:

$$\begin{aligned}&\sum_{k=0}^{K-1}[(A_kg_k)(X'_k)+f_k(X'_k)]\\&=\sum_{k=0}^{K-1}[(A_kg_k)(X'_k)+f_k(X'_k)]I[(P'_k\eta_{k+1})(X'_k)-\eta_k(X'_k)<0]\\&\quad+\sum_{k=0}^{K-1}[(A_kg_k)(X'_k)+f_k(X'_k)]I[(P'_k\eta_{k+1})(X'_k)-\eta_k(X'_k)=0].\end{aligned} \tag{4.16}$$

The first term on the right-hand side is

$$\sum_{k=0}^{K-1}\Big[\frac{(A_kg_k)(X'_k)+f_k(X'_k)}{(P'_k\eta_{k+1})(X'_k)-\eta_k(X'_k)}\Big][(P'_k\eta_{k+1})(X'_k)-\eta_k(X'_k)]$$
$$I[(P'_k\eta_{k+1})(X'_k)-\eta_k(X'_k)<0].$$

Because $|(A_kg_k)(X'_k)+f_k(X'_k)|=|\eta_k(X'_k)|<M_1$ and $|(P'_k\eta_{k+1})(X'_k)-\eta_k(X'_k)|>\varepsilon$, the fraction in the above expression is bounded. From (4.15), we get

$$\lim_{K\to\infty}\frac{1}{K}\sum_{k=0}^{K-1}E'\Big\{[(A_kg_k)(X'_k)+f_k(X'_k)]$$
$$I[(P'_k\eta_{k+1})(X'_k)-\eta_k(X'_k)<0]\Big|X'_0=x\Big\}=0, \tag{4.17}$$

and therefore, from (4.16), it holds that

$$\lim_{K\to\infty}\frac{1}{K}\sum_{k=0}^{K-1}E'\{(A_k g_k)(X'_k)+f_k(X'_k)|X'_0=x\}$$
$$=\lim_{K\to\infty}\frac{1}{K}\sum_{k=0}^{K-1}E'\{[(A_k g_k)(X'_k)+f_k(X'_k)]$$
$$I[(P'_k\eta_{k+1})(X'_k)-\eta_k(X'_k)=0]|X'_0=x\}. \tag{4.18}$$

Next, from Lemma 4.2, $g_k(x)$ is bounded. Similar to (4.17), we have

$$\lim_{K\to\infty}\frac{1}{K}\sum_{k=0}^{K-1}E'\Big\{[(A'_k g_k)(X'_k)+f'_k(X'_k)]$$
$$I[(P'_k\eta_{k+1})(X'_k)-\eta_k(X'_k)<0]\Big|X'_0=x\Big\}=0;$$

and thus,

$$\lim_{K\to\infty}\frac{1}{K}\sum_{k=0}^{K-1}E'\{(A'_k g_k)(X'_k)+f'_k(X'_k)|X'_0=x\}$$
$$=\lim_{K\to\infty}\frac{1}{K}\sum_{k=0}^{K-1}E'\Big\{[(A'_k g_k)(X'_k)+f'_k(X'_k)]$$
$$I[(P'_k\eta_{k+1})(X'_k)-\eta_k(X'_k)=0]\Big|X'_0=x\Big\}. \tag{4.19}$$

Furthermore, condition (b) is equivalent to

$$(A'_k g_k)(x)+f'_k(x)\le(A_k g_k)(x)+f_k(x), \tag{4.20}$$

for all $x\in\mathscr{S}_k$. From (4.20), (4.18), and (4.19), we have

$$\lim_{K\to\infty}\frac{1}{K}E'\Big\{\sum_{k=0}^{K-1}[(A'_k g_k)(X'_{k+1})+f'_k(X'_k)]\Big|X'_0=x\Big\}$$
$$=\lim_{K\to\infty}\frac{1}{K}E'\Big\{\sum_{k=0}^{K-1}\{[(A'_k g_k)(X'_{k+1})+f'_k(X'_k)]$$
$$I[(P'_k\eta_{k+1})(X'_k)-\eta_k(X'_k)=0]\Big|X'_0=x\Big\}$$
$$\le\lim_{K\to\infty}\frac{1}{K}E'\{\sum_{k=0}^{K-1}\Big\{[(A_k g_k)(X'_{k+1})+f_k(X'_k)]$$
$$I[(P'_k\eta_{k+1})(X'_k)-\eta_k(X'_k)=0]\}\Big|X'_0=x\Big\}$$

$$= \lim_{K\to\infty} \frac{1}{K} E'\Big\{ \sum_{k=0}^{K-1} [(A_k g_k)(X'_{k+1}) + f_k(X'_k)] \Big| X'_0 = x \Big\}.$$

From this equation, (4.14), and the performance difference formula (4.12), we conclude that $\eta'(x) \le \eta(x)$ for all $x \in \mathscr{S}_0$. Thus, $\eta' \le \eta$. Similarly, we can prove $\eta'_k \le \eta_k$ for all $k = 0, 1, \ldots$. □

To develop optimality conditions, we note that the long-run average reward is under-selective; thus, the conditions may not need to hold on a non-frequently visited subsequence; with uniformity, it is a sequence of time instants $k_0, k_1, \ldots$, with $\lim_{n\to\infty} \frac{n}{k_n} = 0$.

We use a superscript "$*$" to denote the quantities associated with an optimal policy u^*, e.g., η^*_k, and g^*_k, etc.

Theorem 4.1 (Sufficient optimality conditions, multi-class) *Under Assumption 3.3, a policy $u^* \in \mathscr{D}$ is optimal in the admissible policy space $\mathscr{D}$, if the following conditions are satisfied for all policy $u \in \mathscr{D}$:*

(a) $P^u_k \eta^{u^*}_{k+1} \le \eta^{u^*}_k$, for all $k = 0, 1, \ldots$; and there is an $\varepsilon > 0$ such that if

$$\eta^{u^*}_k(x) - P^u_k \eta^{u^*}_{k+1}(x) > 0, \quad x \in \mathscr{S}_k, \; k = 0, 1, \ldots,$$

then

$$\eta^{u^*}_k(x) - P^u_k \eta^{u^*}_{k+1}(x) > \varepsilon;$$

and

(b) If $(P^u_k \eta^{u*}_{k+1})(x) = \eta^{u^*}_k(x)$, $x \in \mathscr{S}_k$, then

$$(P^u_k g^{u^*}_{k+1})(x) + f^u_k(x) \le (P^{u^*}_k g^{u^*}_{k+1})(x) + f^{u^*}_k(x)$$

holds for all $u \in \mathscr{D}$ and all $k = 0, 1, \ldots$, except on a non-frequently visited subsequence of $k = 0, 1, \ldots$; that is, for any subsequence of $k = 0, 1, \ldots$, denoted by $k_0, k_1, \ldots, k_l, \ldots$, and any sequence of states visited by $\boldsymbol{X}^{u^*}$, $x_{k_0}, x_{k_1}, \ldots, x_{k_l}, \ldots$, $x_{k_l} \in \mathscr{S}_{k_l}$, $l = 0, 1, \ldots$, and sequence of actions α_{k_0}, $\alpha_{k_1}, \ldots, \alpha_{k_l} \in \mathscr{A}_{k_l}(x_{k_l})$, $l = 0, 1, \ldots$, if

$$(P^{\alpha_{k_l}}_{k_l} g^{u^*}_{k_l+1})(x_{k_l}) + f^{\alpha_{k_l}}_{k_l}(x_{k_l}) > (P^{u^*}_{k_l} g^{u^*}_{k_l+1})(x_{k_l}) + f^{u^*}_{k_l}(x_{k_l}),$$

with $(P^{\alpha_{k_l}}_{k_l} \eta^{u^*}_{k_l+1})(x_{k_l}) = \eta^{u^*}_{k_l}(x_{k_l})$, for all $l = 0, 1, \ldots$, then it must hold that

$$\lim_{n\to\infty} \frac{1}{k_n} \sum_{l=1}^{n} \Big\{ (A^{u^*}_{k_l} g^{u^*}_{k_l} + f^{u^*}_{k_l})(x_{k_l}) - (A^{\alpha_{k_l}}_{k_l} g^{u^*}_{k_l} + f^{\alpha_{k_l}}_{k_l})(x_{k_l}) \Big\} = 0.$$

Proof First, we assume that the conditions in (a) and (b) hold for all $k = 0, 1, \ldots$. Then, it follows directly from Lemma 4.3 that $\eta^u \le \eta^{u^*}$ for all $u \in \mathscr{D}$, and u^* is optimal.

Next, if for a policy u, condition (b) holds except for non-frequently visited sequences, then similar to (3.33) in Theorem 3.2, we can prove that

$$\lim_{n\to\infty} \frac{1}{k_n} \sum_{l=1}^{n} E^u \Big\{ \Big[[(A_{k_l}^{u^*} g_{k_l}^{u^*} + f_{k_l}^{u^*})(X_{k_l}^u) - (A_{k_l}^u g_{k_l}^{u^*} + f_{k_l}^u)(X_{k_l}^u)] \Big] \Big| X_0^u = x \Big\} = 0.$$

Then, by the performance difference formula (cf. (4.12)), this equation cannot change the inequality $\eta^u \le \eta^{u^*}$; and thus, u^* is optimal. □

We note that, for simplicity, the theorem is stated with the term "policy", but the conditions can be checked with actions at every state.

The conditions in Theorem 3.1 are not necessary. In fact, we have

Theorem 4.2 (Necessary optimality condition, multi-class) *A policy* $u^* = \{\alpha_0^*, \alpha_1^*, \ldots\} \in \mathscr{D}$ *is optimal for performance (4.13) in the admissible policy space* $\mathscr{D}$*, only if*

$$P_k^{\alpha_k} \eta_{k+1}^{u^*} \le \eta_k^{u^*} \tag{4.21}$$

holds for all decision rules $\alpha_k \in \mathscr{A}_k$ *and all* $k = 0, 1, \ldots$.

Proof Suppose the opposite is true: there is a decision rule α_{k_0}, and at some time instant k_0, (4.21) does not hold; i.e.,

$$P^{\alpha_{k_0}} \eta_{k_0+1}^{u^*} > \eta_{k_0}^{u^*}. \tag{4.22}$$

(The inequality only needs to hold for one component x, with one action $\alpha_{k_0}(x)$, $x \in \mathscr{S}_{k_0}$.) We construct a new policy $\tilde{u} = \{\tilde{\alpha}_0, \tilde{\alpha}_1, \ldots\}$ with $\tilde{\alpha}_{k_0} = \alpha_{k_0}$ and $\tilde{\alpha}_k = \alpha_k^*$ for all $k \neq k_0$, and $\tilde{f}_k = f_k^{u^*}$ for all k. The quantities associated with policy $\tilde{u}$ are denoted by a superscript "$\sim$". The decision rules for both u^* and $\tilde{u}$ are the same for $k \neq k_0$, so $P_k^{u^*} = \tilde{P}_k$ for all $k \neq k_0$, and thus $\tilde{\eta}_{k_0+1} = \eta_{k_0+1}^{u^*}$. By (4.2) and (4.22), we have

$$\tilde{\eta}_{k_0} = P^{\alpha_{k_0}} \tilde{\eta}_{k_0+1} = P^{\alpha_{k_0}} \eta_{k_0+1}^{u^*} > \eta_{k_0}^{u^*}.$$

Then,

$$\tilde{\eta}_{k_0-1} = \tilde{P}_{k_0-1} \tilde{\eta}_{k_0} > P_{k_0-1}^{u^*} \eta_{k_0}^{u^*} = \eta_{k_0-1}^{u^*}.$$

Continuing this process, we get $\tilde{\eta}_0 > \eta_0^{u^*}$, and therefore, $\tilde{u}$ is a better policy, and u^* is not optimal. □

Theorems 4.1 and 4.2 look similar to the results for THMCs, see Puterman (1994), Cao (2007).

4.3 Conclusion

In this chapter, we have shown that in addition to state classification and the performance optimization of single-class TNHMCs, confluencity also plays a fundamental role in the performance optimization of multi-class TNHMCs.

With confluencity and state classification, we can define performance potentials for multi-chain TNHMCs and derive both necessary and sufficient conditions for their long-run average optimal policies. The sufficient conditions do not need to hold for a non-frequently visited time sequence due to the under-selectivity. These results are natural extensions of those for THMCs (Cao 2007; Puterman 1994).

Because in general the "lim inf" performance may not be well defined for multi-class chains, so far in this chapter we have assumed that the "lim" exists in the performance (4.1). This issue will be further discussed in Sect. 4.4.

Just like in Chap. 3, in this chapter, we have applied the relative optimization approach to derive the optimality conditions. The analysis demonstrates the advantages of this approach; see also Cao (2007, 2009, 2015), Cao and Zhang (2008b), Cao et al. (2014), Xia et al. (2014).

4.4 Appendix. The "lim inf" Performance and Asynchronicity

When the limit in (4.1) does not exist, the general form of the average-reward criterion is

$$\eta_k(x) = \liminf_{K\to\infty} \frac{1}{K} E\Big\{ \sum_{l=k}^{k+K-1} f_l(X_l) \Big| X_k = x \Big\}, \tag{4.23}$$

$k = 0, 1, \ldots$. There are some difficulties in using this "lim inf" performance. First, in general, the additive relation (4.2) may not hold. Let us consider an example for some insight.

Example 4.1 (*Cao 2016*) Consider a TNHMC with $\mathscr{S}_k = \{1_k, 2_k\}$ for all $k = 1, 2, \ldots$. Let the transition probabilities be $p(1_{k+1}|2_k) = p(2_{k+1}|1_k) = 1$ and $p(1_{k+1}|1_k) = p(2_{k+1}|2_k) = 0$. Then, there are only two sample paths denoted by $a : 2_1 \to 1_2 \to \ldots$, and $b : 1_1 \to 2_2 \to \ldots$; i.e., the chain has two absorbing classes.

To illustrate (4.2) may not hold for (4.23), we design a case, shown in Table 4.1, by specifically choosing the rewards at different states. The rows marked by $f^a(X_k)$ and $f^b(X_k)$ show the rewards at time k assigned to the states in sample paths a and b, respectively, and

$$\Phi_K^a := \frac{1}{K}\sum_{k=1}^{K} f^a(X_k) \ and\ \Phi_K^b := \frac{1}{K}\sum_{k=1}^{K} f^b(X_k)$$

Table 4.1 An Example in Which the "$liiminf$" Performance is NOT Additive

k:	1	2	3	4	5	6	7	8	9	10	11	12	13	$\cdots$	18	19	$\cdots$	$\cdots$
X_k in a:	1_1	2_2	1_3	2_4	1_5	2_6	1_7	2_8	1_9	2_{10}	1_{11}	2_{12}	1_{13}	$\cdots$	2_{18}	1_{19}	$\cdots$	$\cdots$
$f^a(X_k)$:	1	−1	1	1	−1	−1	1	1	1	1	1	1	−1	$\cdots$	−1	1	$\cdots$	$\cdots$
Φ_k^a:	1	0	$\frac{1}{3}$	$\frac{1}{2}$	$\frac{1}{5}$	0	$\frac{1}{7}$	$\frac{2}{8}$	$\frac{3}{9}$	$\frac{4}{10}$	$\frac{5}{11}$	$\frac{1}{2}$	$\frac{5}{13}$	$\searrow$	0	$\nearrow$	$\frac{1}{2}$	$\cdots$
X_k in b:	2_1	1_2	2_3	1_4	2_5	1_6	2_7	1_8	2_9	1_{10}	2_{11}	1_{12}	2_{13}	$\cdots$	1_{18}	2_{19}	$\cdots$	$\cdots$
$f^b(X_k)$:	−1	1	−1	−1	1	1	−1	−1	−1	−1	−1	−1	1	$\cdots$	1	−1	$\cdots$	$\cdots$
Φ_k^b:	−1	0	$-\frac{1}{3}$	$-\frac{1}{2}$	$-\frac{1}{5}$	0	$-\frac{1}{7}$	$-\frac{2}{8}$	$-\frac{3}{9}$	$-\frac{4}{10}$	$-\frac{5}{11}$	$-\frac{1}{2}$	$-\frac{5}{13}$	$\nearrow$	0	$\searrow$	$-\frac{1}{2}$	$\cdots$

are the corresponding average rewards in the first K steps. As we see, b is a "mirror" case of a, $f^b(X_k) = -f^a(X_k)$, $k = 0, 1, \ldots$. In this example, we have $\liminf_K \Phi_K^a = 0$, $\liminf_K \Phi_K^b = -\frac{1}{2}$, but

$$\liminf_{K\to\infty}(\Phi_K^a + \Phi_K^b) = 0 > \liminf_{K\to\infty} \Phi_K^a + \liminf_{K\to\infty} \Phi_K^b. \qquad (4.24)$$

Thus, the operator "lim inf" is not additive. □

The non-additivity of (4.24) demonstrated in Example 4.1 causes problems in understanding and defining the "lim inf" performance at branching states. Because of it, when x is branching, we may have the undesirable property

$$\eta_k(x) > \sum_{r=1}^{d} p_{k,r}(x)\eta_{\cdot r}. \qquad (4.25)$$

Example 4.2 Suppose that in Example 4.1, there is only one state 1_0 at $k = 0$, $\mathscr{S}_0 = \{1_0\}$, and $f(1_0) = 0$, $p(1_1|1_0) = p(2_1|1_0) = 0.5$. We have $\eta(1_1) = \liminf_{K\to\infty} \Phi_K^a = 0$, and $\eta(2_1) = \liminf_{K\to\infty} \Phi_K^b = -\frac{1}{2}$; but

$$\begin{aligned}
\eta(1_0) &= \liminf_{K\to\infty} \left\{\frac{1}{K}\sum_{k=0}^{K} E[f(X_k)|X_0 = 1_0]\right\} \\
&= \liminf\ (0.5\Phi_K^a + 0.5\Phi_K^b) \\
&= 0 > \liminf\ (0.5\Phi_K^a) + \liminf\ (0.5\Phi_K^b).
\end{aligned}$$

That is, $\eta(1_0) > 0.5\eta(1_1) + 0.5\eta(2_1)$. This confirms (4.25). □

Example 4.1 also indicates that the problem may come from the "synchronization" between the sample paths in two confluent classes. In some sense, it is similar to the difficulty caused by the periodicity in the analysis of steady-state performance in THMCs. We need to exclude the case with such "synchronization" between different confluent classes. Fortunately, such synchronization is very special as shown in the subsequent discussions.

Definition 4.2 Two sequences $\{a_0, a_1, \ldots\}$ and $\{b_0, b_1, \ldots\}$ are said to be *asynchronous* if each of them has an lim-subsequence (Definition 3.1), denoted by $\{a_{k_0}, a_{k_1}, \ldots\}$ and $\{b_{k'_0}, b_{k'_1}, \ldots\}$, respectively, such that for any $K > 0$, there are integers N and N' with $k_N > K$ and $k'_{N'} > K$ and $\{k_N, k_{N+1}, \ldots\} \cap \{k'_{N'}, k'_{N'+1}, \ldots\} \neq \emptyset$.

From the definition, two sequences of time instants are synchronous means that all their lim-subsequences never have the same time instant in the long term.

Lemma 4.4 *(a) If* $\{a_0, a_1, \ldots\}$ *and* $\{b_0, b_1, \ldots\}$ *are asynchronous, then there is a subsequence of integers, denoted by* $k_0, k_1, \ldots$*, such that* a_{k_l} *and* b_{k_l}*,* $l = 0, 1, \ldots$*, are lim-subsequences of* a_k *and* b_k $k = 0, 1, \ldots$*, respectively. We say that these two sequences have a "common" lim-subsequence.*

(b) If $\{a_0, a_1, \ldots\}$ *and* $\{b_0, b_1, \ldots\}$ *are asynchronous, then*

$$\liminf_{n\to\infty} \, (a_n + b_n) = \liminf_{n\to\infty} \, a_n + \liminf_{n\to\infty} \, b_n. \tag{4.26}$$

(c) If $\lim_{n\to\infty} a_n$*, or* $\lim_{n\to\infty} b_n$*, exists, then the two sequences are asynchronous.*

Proof (a) Such a "common" lim-subsequence can be obtained by construction. Let $\{a_{k_0}, a_{k_1}, \ldots\}$ and $\{b_{k'_0}, b_{k'_1}, \ldots\}$ be the two lim-subsequences in Definition 4.2. Suppose that we have constructed the first $l + 1$ elements of the common lim-subsequence up to n_l, denoted by $\{n_0, n_1, \ldots, n_l\}$. Now we set $K = n_l$; then, by the definition of asynchronicity, there are $k_N > K$ and $k'_{N'} > K$, such that $\{k_N, k_{N+1}, \ldots\}$ and $\{k'_{N'}, k'_{N'+1}, \ldots\}$ have a common element, choose it as n_{l+1}. The process goes on and on, leading to a sequence $\{n_0, n_1, \ldots, n_l, \ldots\}$, such that $\{a_{n_l},\ l = 0, 1, \ldots\}$, is a subsequence of $\{a_{k_l},\ l = 0, 1, \ldots\}$; and $\{b_{n_l},\ l = 0, 1, \ldots\}$, is a subsequence of $\{b_{k_l},\ l = 0, 1, \ldots\}$. So both $\lim_{l\to\infty} a_{n_l}$ and $\lim_{l\to\infty} b_{n_l}$ exist.

(b) By (a), we denote the "common" lim-subsequence as $a_{n_0}, a_{n_1} \ldots, a_{n_l}, \ldots$, and $b_{n_0}, b_{n_1} \ldots, b_{n_l}, \ldots, l = 0, 1, \ldots$. Then,

$$\lim_{l\to\infty} (a_{n_l} + b_{n_l}) = \lim_{l\to\infty} a_{n_l} + \lim_{l\to\infty} b_{n_l}.$$

Therefore, by the definition of "lim inf", we have

$$\begin{aligned}\liminf_{k\to\infty} (a_k + b_k) &\leq \lim_{l\to\infty} (a_{n_l} + b_{n_l}) \\ &= \lim_{l\to\infty} a_{n_l} + \lim_{l\to\infty} b_{n_l} \\ &= \liminf_{k\to\infty} \, a_k + \liminf_{k\to\infty} \, b_k.\end{aligned}$$

This, together with the inequality (3.3), leads to (4.26).

(c) If $\lim_{n\to\infty} a_n$ exists, every subsequence of it is a lim-subsequence. The requirement for a_n and b_n, $n = 0, 1, \ldots$, to be asynchronous holds naturally. Then, by part (b), (4.26) holds; this is similar to Lemma 3.1. □

In (4.23), we define

$$\Phi_{k,K}(x) := \frac{1}{K} E\Big\{ \sum_{l=k}^{k+K-1} f_l(X_l) \Big| X_k = x \Big\}, \quad K > k. \tag{4.27}$$

Lemma 4.5 *For any confluent state $x \in \mathscr{R}_{k,r}$, if $\{\Phi_{k,K_l}(x), l = 0, 1, \ldots\}$ is a lim-subsequence of $\{\Phi_{k,K}(x), K = 1, 2, \ldots\}$, then for any other state in the same confluent class, $x' \in \mathscr{R}_{k',r}$, $\{\Phi_{k',K_l}(x'), l = 0, 1, \ldots\}$ is also a lim-subsequence of $\{\Phi_{k',K}(x'), K = 1, 2, \ldots\}$.*

Proof We briefly sketch the proof for $k' = k$. For notational simplicity, let $k = k_0 = 0$. Consider two independent sample paths, denoted by X_l and X'_l, $l = 0, 1, \ldots$, starting from $X_0 = x$ and $X'_0 = x'$, $x, x' \in \mathscr{R}_{0,r}$, respectively. Let $K_0, K_1, \ldots$ be a sequence of integers. With the confluencity, for any $\varepsilon > 0$, we have an L_ε large enough, such that

$$\mathscr{P}(\tau_0(x, x') > K_{L_\varepsilon} | X_0 = x, X'_0 = x') < \varepsilon.$$

Then, for any $K_L > K_{L_\varepsilon}$, $L = 0, 1, \ldots$, we have

$$\begin{aligned}
&\sum_{l=0}^{K_L-1} E\{[f(X'_l) - f(X_l)] | X'_0 = x', X_0 = x\} \\
&= \sum_{l=0}^{K_L-1} \Big\{ E\{[f(X'_l) - f(X_l)] | X'_0 = x', X_0 = x, \tau_0(x, x') \le K_{L_\varepsilon}\} \\
&\qquad \mathscr{P}(\tau_0(x, x') \le K_{L_\varepsilon} | X_0 = x, X'_0 = x') \\
&\qquad + E\{[f(X'_l) - f(X_l)] | X'_0 = x', X_0 = x, \tau_0(x, x') > K_{L_\varepsilon}\} \\
&\qquad \mathscr{P}(\tau_0(x, x') > K_{L_\varepsilon} | X_0 = x, X'_0 = x') \Big\}.
\end{aligned}$$

Similar to the proof of Lemma 3.2, the first term on the right-hand side of the above equation is finite $< 2K_{L_\varepsilon} M_2$. By Assumption 1.1(b) and (3.5), we have

$$\Big| \sum_{l=0}^{K_L-1} E\{[f(X'_l) - f(X_l)] | X'_0 = x', X_0 = x\} \Big| < 2K_L M_2 \varepsilon + 2K_{L_\varepsilon} M_2.$$

Because ε can be any positive number as long as $K_{L_\varepsilon} (< K_L)$ is large enough, we have

$$\lim_{L\to\infty} \frac{1}{K_L} \Big\{ \sum_{l=k}^{k+K_L-1} E\{[f(X'_l) - f(X_l)] | X'_k = x', X_k = x\} \Big\} = 0.$$

This is

$$\lim_{L\to\infty} \{\Phi_{k,K_L}(x') - \Phi_{k,K_L}(x)\} = 0.$$

Because $\lim_{L\to\infty}\{\Phi_{k,K_L}(x)\} = \eta_{\cdot r}$, we have $\lim_{L\to\infty}\{\Phi_{k,K_L}(x')\} = \eta_{\cdot r}$. That means that $\Phi_{k,K_L}(x')$ is a lim-subsequence.

Similarly, we can prove the statement for $k' \neq k$. □

Lemma 4.5 is stronger than that both $\{\Phi_{k,K}(x), K=1, 2, \ldots\}$ and $\{\Phi_{k',K}(x'), K = 1, 2, \ldots\}, x \in \mathscr{R}_{k,r}, x' \in \mathscr{R}_{k',r'}$, are asynchronous (with a common lim-subsequence).

Definition 4.3 Two confluent classes $\mathscr{R}_r$ and $\mathscr{R}_{r'}$, $r, r' = 1, 2, \ldots, d$, are said to be asynchronous to each other, if for any inf-subsequence of $\{\Phi_{0,K}(x), K = 0, 1, \ldots\}$ (Definition 3.1) in $\mathscr{R}_r$, denoted by $\{\Phi_{0,K_l}(x), K_0, K_1, \ldots\}$, there is an inf-subsequence in $\mathscr{R}_{r'}$, $\{\Phi'_{0,K_l}(x'), K_0, K_1, \ldots\}$, such that both subsequences are asynchronous. A TNHMC is said to be asynchronous if all the confluent classes are asynchronous to each other.

Asynchronicity depends on the values of the reward functions f_k, $k = 0, 1, \ldots$. Asynchronicity is not a very strong requirement; it only excludes the case where any two lim-subsequences eventually do not have any common time instants. If the performance limit (4.1) exists, then the Markov chain is asynchronous.

Lemma 4.6 *If $\mathscr{R}_{r_1}$ and $\mathscr{R}_{r_2}$ are asynchronous, and $\mathscr{R}_{r_2}$ and $\mathscr{R}_{r_3}$ are asynchronous, then there is a common lim-subsequence for these three confluent classes; and thus, $\mathscr{R}_{r_1}$ and $\mathscr{R}_{r_3}$ are asynchronous; i.e., asynchronicity of confluent classes is transitive.*

Proof (a) Let $\{\Phi^{r_1}_{k,K}(x_1), K = 1, 2, \ldots\}$ be an inf-subsequence in $\mathscr{R}_{r_1}$. Then, there is an inf-subsequence $\{\Phi^{r_2}_{k,K}(x_2), K = 1, 2, \ldots\}$ in $\mathscr{R}_{r_2}$, (to avoid too many layers of subscripts, we denote them as $\Phi^{r_i}_{k,K}(x_1)$, $i = 1, 2$, but it can be any other subsequences.) and a common lim-subsequence (see Lemma 4.4) $\{\Phi^{r_1}_{k,K_l}(x_1), l = 1, 2, \ldots\}$ and $\{\Phi^{r_2}_{k,K_l}(x_2), l = 1, 2, \ldots\}$. That is,

$$\lim_{l\to\infty} \Phi^{r_i}_{k,K_l}(x_i) = \liminf_{l\to\infty} \Phi^{r_i}_{k,K}(x_i), \quad i = 1, 2. \tag{4.28}$$

Furthermore, $\{\Phi^{r_2}_{k,K_l}(x_2), l = 1, 2, \ldots\}$ is an inf-subsequence; by definition, there is an inf-subsequence $\{\Phi^{r_3}_{k,K_l}(x_3), l = 1, 2, \ldots\}$ in $\mathscr{R}_{r_3}$ which is asynchronous with $\{\Phi^{r_2}_{k,K_l}(x_2), l = 1, 2, \ldots\}$. That is, there is a subsequence of $K_1, K_2, \ldots$, denoted by K_{l_m}, $m = 1, 2, \ldots$, with l_m, $m = 1, 2, \ldots$, being a subsequence of $1, 2, \ldots$, such that $\{\Phi^{r_2}_{k,K_{l_m}}(x_2), m = 1, 2, \ldots\}$ and $\{\Phi^{r_3}_{k,K_{l_m}}(x_3), m = 1, 2, \ldots\}$ are common lim-subsequences. That is,

$$\lim_{m\to\infty} \Phi^{r_3}_{k,K_{l_m}}(x_3) = \liminf_{l\to\infty} \Phi^{r_3}_{k,K_l}(x_3) = \liminf_{K\to\infty} \Phi^{r_3}_{k,K}(x_3).$$

By the above equation and (4.28), $\{\Phi^{r_i}_{k,K_{l_m}}(x_i), m = 1, 2, \ldots\}$, $i = 1, 2, 3$, are the three common lim-subsequences of $\mathscr{R}_{r_1}$, $\mathscr{R}_{r_2}$, and $\mathscr{R}_{r_3}$. In other words, $\lim_{m\to\infty} \Phi^{r_i}_{k,K_{l_m}}(x_i)$, $i = 1, 2, 3$, exist.

By Lemma 4.5, the initial state x_2 and x_3 can be any states in $\mathscr{R}_{r_2}$ and $\mathscr{R}_{r_3}$. The lemma can be easily extended to d confluent classes. $\square$

Consider an asynchronous TNHMC. Let $x_{k,r} \in \mathscr{R}_{k,r}$, $r = 1, 2, \ldots, d$, be the d initial states, each in one confluent class at $k = 0, 1, \ldots$. Let $k_0 (= 0), k_1, \ldots,$ be a common lim-subsequence of $\Phi_{0,K}(x_{0,r})$ defined in (4.27), i.e.,

$$\lim_{n\to\infty} \frac{1}{k_n} E\Big\{ \sum_{n=0}^{k_n-1} f_n(X_n) \Big| X_0 = x_{0,r} \Big\} = \eta_{\cdot r}, \quad r = 1, 2, \ldots, d.$$

With this common inf-sequence, all "lim inf" become "lim". Now, we have

Theorem 4.3 *For an asynchronous TNHMC, Eqs. (4.2) and (4.4) hold.*

Proof Equation (4.4) follows from (4.2). Now we prove (4.2). Let $x \in \mathscr{T}$, and $\{k_0 = 0, k_1, \ldots, k_n, \ldots\}$ be a common lim-subsequence of all the confluent classes. For any $n > N > 0$, we have

$$\Phi_{0,k_n}(x) = \frac{1}{k_n} E\Big\{ \sum_{l=0}^{k_n-1} f_l(X_l) \Big| X_0 = x \Big\}$$
$$= \frac{1}{k_n} E\{ \sum_{l=0}^{k_N-1} f_l(X_l) | X_0 = x\} + \frac{k_n - k_N}{k_n} \frac{1}{k_n - k_N} E\{ \sum_{l=k_N}^{k_n-1} f_l(X_l) | X_0 = x\}, \tag{4.29}$$

and

$$\frac{1}{k_n - k_N} E\{ \sum_{l=k_N}^{k_n-1} f_l(X_l) | X_0 = x\}$$
$$= \frac{1}{k_n - k_N} \Big[\sum_{r=1}^{d} \{ \mathscr{P}(X_{k_N} \in \mathscr{R}_r | X_0 = x) E[\sum_{l=k_N}^{k_n-1} f_l(X_l) | X_N \in \mathscr{R}_r]\}$$
$$+ \mathscr{P}(X_{k_N} \in \mathscr{T} | X_0 = x) E\{ \sum_{l=k_N}^{k_n-1} f_l(X_l) | X_N \in \mathscr{T}\} \Big]. \tag{4.30}$$

By Assumption 1.1b, we have

$$\Big| \frac{1}{k_n - k_N} \Big[\mathscr{P}(X_{k_N} \in \mathscr{T} | X_0 = x) E\{ \sum_{l=k_N}^{k_n-1} f_l(X_l) | X_N \in \mathscr{T}\} \Big] \Big|$$
$$< M_2 \mathscr{P}(X_{k_N} \in \mathscr{T} | X_0 = x).$$

By (2.13), for any $\varepsilon > 0$, there is a $K > 0$ such that $\mathscr{P}(\tau_{0,\mathscr{R}} > K|X_0 = x) < \varepsilon$; i.e., for any N with $k_N > K$, $\mathscr{P}(X_{k_N} \in \mathscr{T}|X_0 = x) < \varepsilon$. In addition, by (2.15), we have $0 < p_{0,r}(x) - \mathscr{P}(X_{k_N} \in \mathscr{R}_r|X_0 = x) < \varepsilon$. Thus, from (4.30), we have

$$\Big|\frac{1}{k_n - k_N} E\{\sum_{l=k_N}^{k_n-1} f_l(X_l)|X_0 = x\} - \sum_{r=1}^{d} p_{0,r}(x)\Big[\frac{1}{k_n - k_N} E\{\sum_{l=k_N}^{k_n-1} f_l(X_l)|X_N \in \mathscr{R}_r\}\Big]\Big| < (1+d)\varepsilon M_2. \tag{4.31}$$

Letting $k_n \to \infty$ in (4.31), we conclude that the limit in the following equation exists and

$$\Big|\lim_{n\to\infty}\Big[\frac{1}{k_n - k_N} E\{\sum_{l=k_N}^{k_n-1} f_l(X_l)|X_0 = x\}\Big] - \sum_{r=1}^{d} p_{0,r}(x)\eta_{\cdot r}\Big| \le (1+d)\varepsilon M_2.$$

From (4.29), this means that $\lim_{n\to\infty} \Phi_{0,k_n}(x)$ exists, and

$$\Big|\lim_{n\to\infty} \Phi_{0,k_n}(x) - \sum_{r=1}^{d} p_{0,r}(x)\eta_{\cdot r}\Big| \le (1+d)\varepsilon M_2.$$

Because ε is any small number, we have

$$\lim_{n\to\infty} \Phi_{0,k_n}(x) = \sum_{r=1}^{d} p_{0,r}(x)\eta_{\cdot r}.$$

By (4.27), we have $\eta_0(x) = \sum_{r=1}^{d} p_{0,r}(x)\eta_{\cdot r}$; similarly, $\eta_k(x) = \sum_{r=1}^{d} p_{k,r}(x)\eta_{\cdot r}$.

Finally, by definition, we have $p_{k,r}(x) = \sum_{y\in\mathscr{S}_{k+1}}[P(y|x)p_{k+1,r}(y)]$, ($p_{k+1,r}(y) = 1$, if $y \in \mathscr{R}_r$). Then, we have

$$\begin{aligned}
&\sum_{y\in\mathscr{S}_{k+1}} [P(y|x)\eta_{k+1}(y)] \\
&= \sum_{y\in\mathscr{S}_{k+1}} [P(y|x)\sum_{r=1}^{d} p_{k+1,r}(y)\eta_{\cdot r}] \\
&= \sum_{r=1}^{d}\{\sum_{y\in\mathscr{S}_{k+1}} [P(y|x)p_{k+1,r}(y)]\eta_{\cdot r}\}
\end{aligned}$$

$$= \sum_{r=1}^{d} p_{k,r}(x)\eta_{\cdot r} = \eta_k(x),$$

and (4.2) follows. □

There may be other technical issues related to the multi-class optimization with the "lim inf" performance criterion that needs to be addressed (Cao 2019), and we will not explore further in the book.

Chapter 5
The Nth-Bias and Blackwell Optimality

In a Markov chain, an Nth bias is the bias of an $(N-1)$th bias, $N = 0, 1, \ldots$, with the 0th bias being the long-run average performance, and 1st bias the standard bias defined in (3.41). The Nth biases, $N \geq 0$, measure the transient performance of the Markov chain at different levels.

In this chapter, we study the optimization problems of the Nth biases for TNHMCs. We derive the optimality conditions for all the Nth biases, $N = 1, 2, \ldots$. We show that the discounted performance can be expressed as a Laurent series in all the Nth biases, the Blackwell optimal policy is a policy that is optimal for all the Nth biases, $N = 0, 1, \ldots$, and the (N + 1)th-bias optimality, is equivalent to the strong N-discount optimality, $N = 0, 1, \ldots$. The results show that the Nth-bias optimality, together with the Blackwell optimality, completely overcomes the under-selectivity issue associated with the long-run average performance. The theory corresponds to the sensitive discount optimality in the literature (Cao 2007; Guo and Hernández-Lerma 2009; Hordijk and Yushkevich 1999a, b; Jasso-Fuentes and Hernández-Lerma 2009a, b; Puterman 1994; Veinott 1969); but there are two distinctions: the Nth-bias optimality approach does not require discounting, and the results presented in this chapter are for TNHMCs. Again, the problem is solved by the relative optimization approach.

In addition to the time-nonhomogeneous nature, the difficulty in extending the Nth-bias optimality theory from THMCs to TNHMCs also lies in determining the spaces of the Nth-bias optimal policies, see the discussions in Sects. 5.2.3 and 5.2.4.

X.-R. Cao, *Foundations of Average-Cost Nonhomogeneous Controlled Markov Chains*, SpringerBriefs in Control, Automation and Robotics,
https://doi.org/10.1007/978-3-030-56678-4_5

5.1 Preliminaries

The long-run average is under-selective; e.g., it does not depend on actions in any finite period. Another related issue is that an optimal long-run average policy may not be optimal for the total reward in any finite period. The latter is true even for THMCs.

To address the issue mentioned above for THMCs, there are roughly three directions in the literature. The first two are basically approximating the long-run average with other more selective criteria. One of them is called the *periodic forecast horizon (PFH) optimality*. In this approach, the long-run average is approximated by the finite period total reward (Bean and Smith 1990; Hopp et al. 1987; Park et al. 1993); and it shows that the PFH optimal policy, as an average optimal policy, is the limit point, defined in some metric, of a sequence of optimal policies for finite horizon problems.

The other approach is the well-known *sensitive discount optimality*. In this approach, the long-run average is approximated by the discount criteria; it includes the n-discount optimality for all integers $n \geq -1$. It is well known that the discount performance converges to the long-run average when the discount factor β goes to one (see, e.g., Cao 2000, Hernández-Lerma and Lasserre 1996). A Blackwell optimal policy optimizes all the discount performance measures with discount factors near one; and it is a long-run average optimal policy with no under-selectivity. A policy is Blackwell optimal if and only if it is n-discount optimal for all $n \geq -1$, see, e.g., Guo and Hernández-Lerma (2009), Hordijk and Yushkevich (1999a, b), Jasso-Fuentes and Hernández-Lerma (2009a), Puterman (1994), Veinott (1969).

The third approach is the Nth-bias approach, which works directly on an infinite horizon with no discounting. It starts with the bias optimality (e.g., Guo et al. 2009; Lewis and Puterman 2001; Puterman 1994; Jasso-Fuentes and Hernández-Lerma 2009a). The $(N+1)$th bias is the bias of the Nth bias; $N \geq 1$. The Nth biases, $N = 1, 2, \ldots$, describe the transient behavior of a Markov chain at different levels. The Nth-bias optimality conditions were derived in Cao and Zhang (2008b) and Cao (2007) for homogeneous discrete-time finite Markov chains and in Guo and Hernández-Lerma (2009) and Zhang and Cao (2009) for homogeneous continuous-time chains. The policy that optimizes all the Nth biases, $N = 0, 1, \ldots$, is the one with no under-selectivity. It is shown that this approach is equivalent to the sensitive discount optimality; in particular, the discounted performance can be expanded into a Laurent series in the Nth biases; the policy that optimizes all the Nth biases is the Blackwell optimal policy, and vice versa.

In this chapter, we extend the theory of Nth-bias optimality and Blackwell optimality to TNHMCs. The necessary and sufficient conditions for the Nth-bias optimal and Blackwell optimal policies are derived. In addition, we derive the Laurent series in Nth biases and prove that a policy is Blackwell optimal if and only if it is Nth-bias optimal for all $N \geq 0$. For simplicity, we assume that no states are joining the chain at time $k > 0$; i.e., $\mathscr{S}_{k,out} = \mathscr{S}_{k+1}$, $k = 0, 1, \ldots$ (cf. (1.5)), and we only present the results for uni-chains.

To study the Nth biases, we need an assumption stronger than Assumption 3.4:

Assumption 5.7 (*Geometrically ergodic Chan 1989; Meyn and Tweedie 2009*) There exist a $1 > q > 0$ and a constant C such that for all $x \in \mathscr{S}_k$, $k = 1, 2, \ldots$, and $l > 0$, it holds that

$$E\{|f_{l+k}(X_{l+k}) - \eta| \, |X_k = x\} < Cq^l.$$

This assumption holds for THMCs, with $P_k \equiv P$, for all $k = 0, 1, \ldots$. For more discussion on geometrical ergodicity, see Sect. 5.3.4.

Definition 5.1 A policy $u = (\mathbb{P}, \not{f})$ is said to be *admissible*, if (a) Assumptions 1.1, 1.2, and 5.7 hold, and (b) $\boldsymbol{X}^u$ satisfies the strong confluencity. The space of all the admissible policies is denoted by $\mathscr{D}$.

We refer to the bias $\hat{g}_k(x)$ in the form of (3.41) as the first-order bias and denote it as $g_k^{(1)}(x)$, $x \in \mathscr{S}_k$, $k = 0, 1, \ldots$. Likewise, we denote the relative potential (3.6) as $\gamma_k^{(1)}(x, y)$, $x, y \in \mathscr{S}_k$, $k = 0, 1, \ldots$. We wish to optimize the bias while keeping the average-reward optimal; i.e., to find a $u^* \in \mathscr{D}_0$ such that

$$g_k^{(1),u^*}(x) = \max_{u \in \mathscr{D}_0} \{g_k^{(1),u}(x)\}, \tag{5.1}$$

for all $x \in \mathscr{S}_k$, $k = 0, 1, \ldots$, where $\mathscr{D}_0 \subseteq \mathscr{D}$ is *the space of all average-reward optimal policies*, which is determined in Sect. 3.4.1.2.

Let u^* be an optimal long-run average policy, and $\mathscr{C}_0$ be the policy space determined by (cf. (3.46))

$$\begin{aligned}\mathscr{A}_{0,k}(x) := \Big\{\alpha \in \mathscr{A}_k(x) : \sum_y P_k^\alpha(y|x) g_{k+1}^{u^*}(y) + f_k^\alpha(x) \\ = \sum_y P_k^{u^*}(y|x) g_{k+1}^{u^*}(y) + f_k^{u^*}(x)\Big\}, \quad x \in \mathscr{S}_k;\end{aligned}$$

and (see (3.47) and (3.48))

$$\mathscr{C}_{0,k} := \prod_{x \in \mathscr{S}_k} \mathscr{A}_{0,k}(x), \quad \mathscr{C}_0 := \prod_{k=0}^{\infty} \mathscr{C}_{0,k},$$

where $\mathscr{A}_{0,k}(x)$ is the set of long-run average optimal actions at state x and time k.

By Lemma 3.12, all the bias optimal policies are in fact in the space $\mathscr{C}_0$, and thus the problem (5.1) becomes

$$g_k^{(1),u^*}(x) = \max_{u \in \mathscr{C}_0} \{g_k^{(1),u}(x)\},$$

for all $x \in \mathscr{S}_k$, $k = 0, 1, \ldots$.

With the terminology in the Nth-bias optimality, the relative bias-potential, defined by $\chi_k(x, y)$ in (3.52), is called the *relative 2nd-order bias*, and for any $x, y \in \mathscr{S}_k$, it is defined by

$$\gamma_k^{(2)}(x, y) := E\Big\{ \sum_{l=k}^{k+\tau_k(x,y)} [g_l^{(1)}(X_l) - g_l^{(1)}(X_l')] \Big| X_k' = y, X_k = x \Big\}, \quad k = 0, 1, \ldots,$$

where X_l and X_l', $l = 0, 1, \ldots$, are two independent sample paths under the same policy with two different initial states x and y, respectively. Similar to the $w_k(x)$ in (3.53), there is a bias of bias, or the $2nd$ bias, $g_k^{(2)}(x)$, $x \in \mathscr{S}_k$, such that

$$\gamma_k^{(2)}(x, y) = g_k^{(2)}(y) - g_k^{(2)}(x), \quad x, y \in \mathscr{S}_k, \quad k = 0, 1, \ldots.$$

Let $g_k^{(2)} = (g_k^{(2)}(1), \ldots, g_k^{(2)}(S_k))^T$; then, it satisfies the Poisson equation for the 2nd bias (cf. (3.55)). The conditions for bias optimal policies are stated in Theorem 3.5.

5.2 The Nth-Bias Optimality

5.2.1 The Nth Bias and Main Properties

The long-run average is defined in (1.7) by

$$\eta_k(x) = \lim_{L\to\infty} \frac{1}{L} E\Big\{ \sum_{l=0}^{L-1} f_{k+l}(X_{k+l}) \Big| X_k = x \Big\}, \quad x \in \mathscr{S}_k, \ k = 0, 1, \ldots. \qquad (5.2)$$

If all the states are confluent to each other (uni-chain), we have $\eta_k(x) \equiv \eta$ for all $x \in \mathscr{S}_k$, $k = 0, 1, \ldots$. From (5.2), it holds that

$$\lim_{L\to\infty} E\{[f_{l+k}(X_K) - \eta]|X_k = x\} = 0.$$

By Assumption 5.7, it is easy to verify that the bias

$$g_k^{(1)}(x) \equiv g_k(x) := \lim_{K\to\infty} E\{\sum_{l=0}^{K} [f_{k+l}(X_{k+l}) - \eta]|X_k = x\}. \qquad (5.3)$$

exists, and $g_k^{(1)} = (g_k^{(1)}(1), \ldots, g_k^{(1)}(S_k))^T$ satisfies the Poisson equation (cf. (3.42))

$$A_k g_k^{(1)} + f_k = \eta e. \qquad (5.4)$$

We also refer to η as the 0th bias, and denote it as $g^{(0)} := \eta e$.

By Assumption 5.7, we have

$$|E[g_k^{(1)}(X_k)|X_0 = i]| \leq Cq^k + Cq^{k+1} + \cdots = C_1 q^k, \tag{5.5}$$

where $C_1 = \frac{C}{1-q}$. So,

$$\lim_{K\to\infty} E\{g_K^{(1)}(X_K)|X_0 = x\} = 0.$$

As shown in Theorem 3.5, with the second-order (the 2nd) bias $g_k^{(2)}$, we may find the optimal first-order (the 1st) bias optimal policy, which measures the transient performance. However, the bias still has a bias, which measures the transient performance at a higher level, and it might be desirable to maximize this 2nd-order bias, $g_k^{(2)}(x), x \in \mathscr{S}_k, k = 0, 1, \ldots$. Likewise, we may need to optimize even higher order transient performance, i.e., the bias of 2nd bias, etc.

We define the Nth bias recursively. The Nth bias at time k, $g_k^{(N)}, k = 0, 1, \ldots$, is defined by the $(N-1)$th bias via the *Nth-bias Poisson equation*, $N \geq 2$:

$$g_k^{(N)} - P_k g_{k+1}^{(N)} = -g_k^{(N-1)}, \tag{5.6}$$

with $g_k^{(1)} := g_k$ being the 1st bias defined in (5.3). We may write (5.6) as

$$A_k g_k^{(N)} - g_k^{(N-1)} = 0. \tag{5.7}$$

The equation can only determine $g_k^{(N)}$ up to an additive constant.

In particular, for the 2nd bias, it is

$$A_k g_k^{(2)} - g_k^{(1)} = 0. \tag{5.8}$$

Note that in (5.4) it holds that $\lim_{K\to\infty} E[f_K(X_K)|X_0 = x] = \eta$, and in (5.8) it holds that $\lim_{K\to\infty} E[g_K^{(1)}(X_K)|X_0 = x] = 0$. Thus, the Poisson equation (5.8) takes the same form as (5.4), with $g_k^{(1)}$ corresponding to f_k. Hence $g_k^{(2)}$ is indeed the "bias of bias."

Similar to (3.52), the Nth bias can also be determined via the *relative Nth-bias* for any $x, y \in \mathscr{S}_k$, defined by

$$\gamma_k^{(N)}(x, y) := E\Big\{ \sum_{l=k}^{k+\tau_k(x,y)} [g_l^{(N-1)}(X_l) - g_l^{(N-1)}(X_l')] \Big| X_k' = y, X_k = x \Big\}, \quad k = 0, 1, \ldots, N \geq 2, \tag{5.9}$$

where X_l and X_l', $l = 0, 1, \ldots$, are two independent sample paths under the same policy with two different initial states x and y, respectively. By the strong confluencity and Assumption 1.1, we have $|\gamma_k^{(N)}(x, y)| < M < \infty$.

Now, we derive the Poisson equation, starting from (5.9). For any x, we have $\tau_k(x,x)=0$ and $\gamma_k^{(N)}(x,x)=0$. If $x \neq y$, then $\tau_k(x,y)>0$, and from (5.9), we have (cf. (3.8))

$$\begin{aligned}\gamma_k^{(N)}(x,y) = {} & [g_k^{(N-1)}(y) - g_k^{(N-1)}(x)] \\ & + \sum_{x' \in \mathscr{S}_{k+1}} \sum_{y' \in \mathscr{S}_{k+1}} \gamma_{k+1}^{(N)}(x',y') P_k(x'|x) P_k(y'|y). \end{aligned} \tag{5.10}$$

Let $\Gamma_k^{(N)} = [\gamma_k^{(N)}(x,y)]_{x,y \in \mathscr{S}_k}$, $e = (1,\dots,1)^T$, $g_k^{(N-1)} := (g_k^{(N-1)}(1),\dots, g_k^{(N-1)}(S_k))^T$ and $G_k^{(N-1)} := e[g_k^{(N-1)}]^T - g_k^{(N-1)} e^T$. Then, (5.10) takes the form

$$\Gamma_k^{(N)} - P_k \Gamma_{k+1}^{(N)} P_k^T = G_k^{(N-1)}, \quad k = 0, 1, \dots, \tag{5.11}$$

for $N \geq 1$, with $G_k^{(0)} := e[f_k]^T - f_k e^T$, $f_k := (f_k(1),\dots,f_k(S_k))^T$.

Next, similar to (3.13) and based on (5.9), there is an Nth bias, $g_k^{(N)}(x)$, such that

$$\gamma_k^{(N)}(x,y) = g_k^{(N)}(y) - g_k^{(N)}(x), \quad x, y \in \mathscr{S}_k, \; k = 0, 1, \dots.$$

Let $g_k^{(N)} = (g_k^{(N)}(1),\dots,g_k^{(N)}(S_k))^T$. With (5.11), we may verify that $g_k^{(N)}$ indeed satisfies the Poisson equation (5.6) (up to an additive constant).

Repeatedly using (5.6), we have

$$\begin{aligned} g_k^{(N)} &= -g_k^{(N-1)} + P_k g_{k+1}^{(N)} \\ &= -g_k^{(N-1)} + P_k[-g_{k+1}^{(N-1)} + P_k g_{k+1}^{(N)}] \\ &= -g_k^{(N-1)} - P_k g_{k+1}^{(N-1)} + P_k P_{k+1} g_{k+2}^{(N)} \\ &= \cdots\cdots \\ &= -g_k^{(N-1)} - \sum_{l=k}^{K-1} \{[\prod_{m=k}^{l} P_m] g_{l+1}^{(N-1)}\} + [\prod_{l=k}^{K} P_l] g_{K+1}^{(N)}. \end{aligned}$$

Componentwise, this is

$$\begin{aligned} g_k^{(N)}(x) = {} & -\sum_{l=k}^{K} E[g_l^{(N-1)}(X_l)|X_k = x] \\ & + E[g_{K+1}^{(N)}(X_{K+1})|X_k = x]. \end{aligned} \tag{5.12}$$

Next, if the first term on the right-hand side of (5.12) converges, i.e.,

$$\left| \sum_{l=k}^{\infty} E[g_l^{(N-1)}(X_l)|X_k = x] \right| < C < \infty, \tag{5.13}$$

then because $g_k^{(N)}$ is only up to an additive constant, we may take

$$g_k^{(N)}(x) = -\sum_{l=k}^{\infty} E[g_l^{(N-1)}(X_l)|X_k = x], \ N \geq 1. \tag{5.14}$$

as the Nth bias. It can be easily verified that it satisfies the Nth-bias Poisson equation (5.7). With (5.12) and (5.14), we have

$$\lim_{K\to\infty} E[g_K^{(N)}(X_K)|X_k = x] = 0, \ \ \forall x \in \mathscr{S}_k, \ N \geq 1. \tag{5.15}$$

We may prove recursively that with the geometrical ergodicity (Assumption 5.7), the first term on the right-hand side of (5.12) indeed converges as $K \to \infty$, for all N and x. First, for $N = 2$, $k = 0$, by (5.5), we have

$$\begin{aligned}&\sum_{l=0}^{\infty} E[|g_l^{(1)}(X_l)||X_0 = x]\\ \leq& C_1 + C_1 q + C_1 q^2 + \cdots = \frac{C_1}{1-q} = C_2,\end{aligned}$$

where $C_2 = \frac{C_1}{1-q}$. Thus, the statement holds for $N = 2$, $k = 0$, and similarly, it holds for all $k = 0, 1, \ldots$. So, (5.14) holds for $g_k^{(2)}(x)$, for all k. From this, it holds that

$$\begin{aligned}&|E[g_k^{(2)}(X_k)|X_0 = x]|\\ \leq& C_1 q^k + C_1 q^{k+1} + C_1 q^{k+2} + \cdots = \frac{C_1 q^k}{1-q} = C_2 q^k.\end{aligned}$$

From this and by Assumption 5.7, we may prove the results for $N = 3$. Recursively, we may prove (5.13) and (5.15) for all N; in particular, we have

$$|E[g_{k+l}^{(N)}(X_k)|X_k = x]| \leq C_N q^l. \tag{5.16}$$

Next, from (5.3) and (5.14), we have

$$g_0^{(2)}(x) = -\sum_{k=0}^{\infty} \{(k+1) E\{[f_k(X_k) - \eta]|X_0 = x\},$$

or

$$g_0^{(2)}(x) = -\sum_{k=0}^{\infty} \binom{k+1}{1} E\left\{[f_k(X_k) - \eta]|X_0 = x\right\}.$$

By

$$\binom{k+N}{N} = \binom{k+N-1}{N-1} + \binom{k+N-2}{N-1} + \cdots + \binom{N-1}{N-1},$$

we may recursively prove that for all $N > 0$,

$$g_0^{(N+1)}(x) = (-1)^N \sum_{k=0}^{\infty} \binom{N+k}{N} E\left\{[f_k(X_k) - \eta] | X_0 = x\right\}. \tag{5.17}$$

This is the same as the equation for THMCs (cf. (4.81) in Cao 2007).

5.2.2 The Nth-Bias Difference Formula

As in the bias case, we wish to optimize the $(N+1)$th bias (in the form of (5.17) and (5.14)) in the Nth-bias optimal policy space, denoted by $\mathscr{D}_N$, $N = 1, 2, \ldots$, $\mathscr{D}_N \subseteq \mathscr{D}_{N-1}$, with $\mathscr{D}_0 \subseteq \mathscr{D}$ being the long-run average optimal space. An $(N+1)$th-bias optimal policy, denoted by $u^{(N+1),*}$, is defined by

$$g_k^{(N+1),u^{(N+1),*}} = \max_{u \in \mathscr{D}_N} \{g_k^{(N+1),u}\} =: g_k^{(N+1),*}, \tag{5.18}$$

for all $k = 0, 1, \ldots$.

We first derive the Nth-bias difference formula for all $N \geq 0$.

Lemma 5.1 *Consider two policies $u, u' \in \mathscr{D}$, for which Assumptions 1.1 and 5.7 hold. Let X_k^u, $X_k^{u'}$, $g_k^{(N),u}$, $g_k^{(N),u'}$, $k = 0, 1, \ldots$, be the corresponding Markov chains and their Nth biases. If $g_k^{(N),u} = g_k^{(N),u'}$, $k = 0, 1, \ldots$, then for $N = 0$, we have*

$$\begin{aligned} & g_0^{(1),u'}(x) - g_0^{(1),u}(x) \\ = & \lim_{K\to\infty} \frac{1}{K} \sum_{k=0}^{K-1} E^{u'}\left\{[A_k^{u'} - A_k^u] g_k^{(2),u}(X_k^{u'}) \Big| X_0^{u'} = x\right\} \\ & + \lim_{K\to\infty} \frac{1}{K} \sum_{l=1}^{K-1} (K-l-1) E^{u'}\Big\{(P_{l-1}^{u'} g_l^{(1),u} + f_{l-1}^{u'})(X_l^h) \\ & \qquad - (P_{l-1}^{u} g_l^{(1),u} + f_{l-1}^{u})(X_l^{u'}) \Big| X_0^{u'} = x\Big\}; \end{aligned} \tag{5.19}$$

and for $N \geq 1$, we have

$$\begin{aligned} & g_0^{(N+1),u'}(x) - g_0^{(N+1),u}(x) \\ = & \lim_{K\to\infty} \frac{1}{K} \sum_{k=0}^{K-1} E^{u'}\left\{[A_k^{u'} - A_k^u] g_k^{(N+2),u}(X_k^{u'}) \Big| X_0^{u'} = x\right\} \\ & + \lim_{K\to\infty} \frac{1}{K} \sum_{l=1}^{K-1} (K-l-1) E^{u'}\Big\{[A_{l-1}^{u'} - A_{l-1}^u] \\ & \qquad g_{l-1}^{(N+1),u}(X_{l-1}^{u'}) \Big| X_0^{u'} = x\Big\}. \end{aligned} \tag{5.20}$$

Proof For $N = 0$, the difference formula (5.19) is the same as (3.60).

Let $N \ge 1$. By the Poisson equation (5.6), for any policy u, we have

$$g_k^{(N+1),u} - P_k g_{k+1}^{(N+1),u} = -g_k^{(N),u}.$$

Therefore, if $g_k^{(N),u'} = g_k^{(N),u}$, we have

$$\begin{aligned} & g_0^{(N+1),u'} - g_0^{(N+1),u} \\ =& \{P_0^{u'} g_1^{(N+1),u'} - g_0^{(N),u'}\} - \{P_0^u g_1^{(N+1),u} - g_0^{(N),u}\} \\ =& [P_0^{u'} - P_0^u] g_1^{(N+1),u} + P_0^{u'}[g_1^{(N+1),u'} - g_1^{(N+1),u}] \\ =& [P_0^{u'} - P_0^u] g_1^{(N+1),u} + P_0^{u'}[P_1^{u'} - P_1^u] g_2^{(N+1),u} \\ & + P_0^{u'} P_1^{u'} [g_2^{(N+1),u'} - g_2^{(N+1),u}]. \end{aligned}$$

Recursively, we get

$$\begin{aligned} & g_0^{(N+1),u'} - g_0^{(N+1),u} \\ = & \sum_{l=1}^{k} \left\{[\prod_{i=0}^{l-2} P_i^{u'}][P_{l-1}^{u'} - P_{l-1}^u] g_l^{(N+1),u}\right\} + [\prod_{i=0}^{k-1} P_i^{u'}][g_k^{(N+1),u'} - g_k^{(N+1),u}], \end{aligned}$$

in which by convention $\prod_{i=0}^{-1} P_i^{u'} = I$. Componentwise, for any $x \in \mathscr{S}_0$ it holds that

$$\begin{aligned} & g_0^{(N+1),u'}(x) - g_0^{(N+1),u}(x) \\ =& E^{u'}\{g_k^{(N+1),u'}(X_k^{u'}) - g_k^{(N+1),u}(X_k^{u'}) | X_0^{u'} = x\} \\ & + \sum_{l=1}^{k} E^{u'}\left\{[P_{l-1}^{u'} - P_{l-1}^u] g_l^{(N+1),u}(X_l^{u'}) \Big| X_0^{u'} = x\right\}, \quad k \ge 0. \end{aligned}$$

Averaging it over k from 0 to $K-1$ and taking the limit, we have

$$\begin{aligned} & g_0^{(N+1),u'}(x) - g_0^{(N+1),u}(x) \\ =& \frac{1}{K} \lim_{K\to\infty} \sum_{k=0}^{K-1} E^{u'}\left\{ g_k^{(N+1),u'}(X_k^{u'}) - g_k^{(N+1),u}(X_k^{u'}) \Big| X_0^{u'} = x \right\} \\ & + \frac{1}{K} \lim_{K\to\infty} \sum_{k=0}^{K-1} \sum_{l=1}^{k} E^{u'}\left\{ [P_{l-1}^{u'} - P_{l-1}^{u}] g_l^{(N+1),u}(X_l^h) \Big| X_0^{u'} = x \right\}. \end{aligned} \tag{5.21}$$

Next, by the Poisson equation (5.7) we have

$$\begin{aligned} & \lim_{K\to\infty} \frac{1}{K} \sum_{k=0}^{K-1} E^{u'}\left\{ g_k^{(N+1),u'}(X_k^{u'}) - g_k^{(N+1),u}(X_k^{u'}) \Big| X_0^{u'} = x \right\} \\ = & \lim_{K\to\infty} \frac{1}{K} \sum_{k=0}^{K-1} E^{u'}\left\{ A_k^{u'} g_k^{(N+2),u}(X_k^{u'}) - A_k^{u} g_k^{(N+2),u}(X_k^{u'}) \Big| X_0^{u'} = x \right\} \\ & + \lim_{K\to\infty} \frac{1}{K} \sum_{k=0}^{K-1} E^{u'}\left\{ A_k^{u'} g_k^{(N+2),u'}(X_k^{u'}) - A_k^{u'} g_k^{(N+2),u}(X_k^{u'}) \Big| X_0^{u'} = x \right\}. \end{aligned} \tag{5.22}$$

Applying Dynkin's formula, we get

$$\begin{aligned} & \sum_{k=0}^{K-1} E^{u'}\left\{ A_k^{u'} g_k^{(N+2),u'}(X_k^{u'}) - A_k^{u'} g_k^{(N+2),u}(X_k^{u'}) \Big| X_0^{u'} = x \right\} \\ =& E^{u'}\left\{ g_K^{(N+2),u'}(X_K^{u'}) - g_K^{(N+2),u}(X_K^{u'}) \Big| X_0^{u'} = x \right\} \\ & - [g_0^{(N+2),u'}(X_0^{u'}) - g_0^{(N+2),u}(X_0^{u'})]. \end{aligned}$$

Because, under Assumption 5.7, the Nth bias is bounded for any N (cf. (5.13) and (5.14)), from the above equation, we have

$$\begin{aligned} & \lim_{K\to\infty} \frac{1}{K} \sum_{k=0}^{K-1} E^{u'}\left\{ A_k^{u'} g_k^{(N+2),u'}(X_k^{u'}) \right. \\ & \left. - A_k^{u'} g_k^{(N+2),u}(X_k^{u'}) \Big| X_0^{u'} = x \right\} = 0. \end{aligned} \tag{5.23}$$

In addition, by (3.22), we have

$$[P_{l-1}^{u'} - P_{l-1}^{u}] g_l^{(N+1),u} = [A_{l-1}^{u'} - A_{l-1}^{u}] g_{l-1}^{(N+1),u}, \tag{5.24}$$

for all $l \geq 1$. Finally, by (5.22)–(5.24), from (5.21), we have

$$
\begin{aligned}
& g_0^{(N+1),u'}(x) - g_0^{(N+1),u}(x) \\
= & \lim_{K\to\infty} \frac{1}{K} \sum_{k=0}^{K-1} E^{u'} \Big\{ [A_k^{u'} - A_k^{u}] g_k^{(N+2),u}(X_k^{u'}) \Big| X_0^{u'} = x \Big\} \\
& + \frac{1}{K} \lim_{K\to\infty} \sum_{k=0}^{K-1} \sum_{l=1}^{k} E^{u'} \Big\{ [A_{l-1}^{u'} - A_{l-1}^{u}] \\
& \qquad g_{l-1}^{(N+1),u}(X_{l-1}^{u'}) \Big| X_0^{u'} = x \Big\}.
\end{aligned}
$$

Exchanging the order of the two summations over k and l in the last term of the above equation, we get the *Nth-bias difference formula* (5.20). □

5.2.3 The Nth-Bias Optimality Conditions I

By the Nth-bias difference formula (5.20), it is clear that under Assumptions 1.2 and 5.7, an Nth-bias optimal policy $u^* \in \mathscr{D}_N$ is $(N+1)$th-bias optimal, if

(a) The biases $g_k^{(N+1),u^*}$, $k = 0, 1, \ldots$, $u^* \in \mathscr{D}_N$, satisfy

$$
A_k^{u^*} g_k^{(N+1),u^*}(x) = \max_{u \in \mathscr{D}_N} \{ A_k^{u} g_k^{(N+1),u^*}(x) \}, \quad x \in \mathscr{S}_k,
$$

and

(b) The biases $g_k^{(N+2),u^*}$, $k = 0, 1, \ldots$, $u^* \in \mathscr{D}_N$, satisfy

$$
(A_k^{u^*} g_k^{(N+2),u^*})(x) = \max_{u \in \mathscr{D}_N} \{ (A_k^{\alpha} g_k^{(N+2),u^*})(x) \}, \quad x \in \mathscr{S}_k.
$$

However, there are two issues left with the above statement: (a) Existence: Do these two conditions contradict each other? and (b) Feasibility: what is the structure of the searching space $\mathscr{D}_N$? in other words, is it a Cartesian product of the action spaces so that the search can be implemented in the action spaces? (It is impossible to search u^* in the policy space because of its huge dimension.)

To address these issues, we first need to determine the space of all the Nth-bias optimal policies, $\mathscr{D}_N$, $N = 0, 1, \ldots$. The $(N+1)$th-bias optimal conditions have to be verified in the space $\mathscr{D}_N$. In general, the Nth-bias optimal policy space $\mathscr{D}_N$ is determined in terms of the Nth-bias optimal conditions, which in turn depend on the $(N-1)$th-bias optimal policy space $\mathscr{D}_{N-1}$. Therefore, we have to fulfill the tasks recursively: From the $(N-1)$th-bias optimal policy space $\mathscr{D}_{N-1}$, we determine the Nth-bias optimal conditions, from which, we determine the Nth-bias optimal policy space $\mathscr{D}_N$. We have done it for $N = 1$ in Chap. 3, so we start with $N > 1$, and the derivation and results are stated for all N in general, and they have to be understood in a recursive way.

Similar to Lemma 3.4, we have

Lemma 5.2 *Suppose Assumptions 1.1 and 5.7 hold for a Markov chain* $\boldsymbol{X} = \{X_k, k = 0, 1, \ldots\}$. *Let its* $(N-1)$*th bias be* $g_k^{(N-1)}$, $N > 1$. *If* $g_k^{(N)}$ *and* $g_k'^{(N)}$ *satisfy the Poisson equation (5.7) (with the same* A_k *and* $g_k^{(N-1)}$*) for all* $k = 0, 1, \ldots$, *then*

$$g_k'^{(N)} - g_k^{(N)} = ce,$$

with c *being a constant,* $k = 0, 1, \ldots$, $N \geq 0$.

When $N = 1$, *the* 0*th bias is the long-run average* η. *Then, if* $g_k^{(1)}$, *and* $g_k'^{(1)}$ *satisfy the Poisson equation (3.42) (with the same* A_k, f_k, *and* η*), then* $g_k'^{(N)} - g_k^{(N)} = ce$, *with* c *being a constant,* $k = 0, 1, \ldots$.

Proof We prove the case with $N > 1$ (the case with $N = 1$ is shown in Lemma 3.4). Both $g_k^{(N)}$ and $g_k'^{(N)}$ satisfy the same Poisson equation (5.7), so, $A_k g_k'^{(N)} = A_k g_k^{(N)} = g_k^{(N-1)}$; thus, we have

$$\begin{aligned}
&[g_k'^{(N)} - g_k^{(N)}](x) = P_k[g_{k+1}'^{(N)} - g_{k+1}^{(N)}](x) = \cdots\cdots \\
&= \lim_{K\to\infty} E\{[g_K'^{(N)} - g_K^{(N)}](X_K)|X_k = x\}.
\end{aligned}$$

By (5.15), we have $[g_k'^{(N)} - g_k^{(N)}](x) = \lim_{K\to\infty} E\{[g_K'^{(N)}](X_K)|X_k = x\}$; in particular, this limit exists. By the confluencity, this does not depend on x and k. The lemma is proved. □

As in Lemma 3.12, for necessary conditions, we need a uniformity condition. Similar to (3.62)–(3.64), for any $N > 0$, let u^* be an Nth-bias optimal policy, we define

$$\Delta_k^{(N)}(\alpha, x) := (A_k^\alpha g_k^{(N),u^*})(x) - (A_k^{u^*} g_k^{(N),u^*})(x).$$

Set

$$\Lambda_k^{(N)} := \{\Delta_k^{(N)}(\alpha, x) > 0 : \ all\ \alpha \in \mathscr{A}_k(x),\ x \in \mathscr{S}_k\}, \tag{5.25}$$

and

$$\Lambda^{(N)} := \cup_{k=0}^{\infty} \Lambda_k^{(N)}.$$

Later we will see that $\alpha \in \mathscr{A}_k(x)$ in (5.25) may be replaced by a smaller space, $\alpha \in \mathscr{A}_{N,k}(x)$ (see (5.37)).

Assumption 5.8 $\Delta_k^{(N)}(\alpha, x) > 0$ uniformly on $\Lambda^{(N)}$ for every Nth-bias optimal policy u^*.

Now, we have the following necessary condition.

Theorem 5.1 (Necessary Condition) *Suppose Assumption 5.8 holds for the* $(N+1)$*th bias, i.e.,* $\Delta_k^{(N+1)}(\alpha, x) > 0$, $N > 0$, *uniformly on* $\Lambda^{(N)}$. *A necessary condition for an* N*th-bias optimal policy* $u^* \in \mathscr{D}_N$ *to be* $(N+1)$*th bias optimal is that*

$$(g_k^{(N),*} =)\ A_k^{u^*} g_k^{(N+1),u^*} = \max_{u \in \mathscr{D}_N} [A_k^u g_k^{(N+1),u^*}] \tag{5.26}$$

holds for all $k = 0, 1, \ldots$. *In other words,* $u^* \in \bar{\mathscr{D}}_N$, *with*

$$\bar{\mathscr{D}}_N := \{u^* \in \mathscr{D}_N : A_k^{u^*} g_k^{(N+1),u^*} = \max_{u \in \mathscr{D}_N} [A_k^u g_k^{(N+1),u^*}]\} \tag{5.27}$$

denoting the space of all the policies satisfying (5.26) (with $\bar{\mathscr{D}}_0 = \mathscr{C}_0$*).*

Proof The proof is similar to that of Lemma 3.12. Let $u \in \mathscr{D}_N - \bar{\mathscr{D}}_N$ be an Nth-bias optimal policy, but (5.26) does not hold for all k. By Theorem 5.2.b below (in a recursive way), with Assumption 5.8, (5.26) holds except for a non-frequently visited sequence $k_0, k_1, \ldots$. More precisely, we denote $u = \{\alpha_0, \alpha_1, \ldots\}$, with α_k being the decision rule at time k; then, the condition

$$A_k^{\alpha_k} g_k^{(N+1),u} = \max_{u' \in \mathscr{D}_N} [A_k^{u'} g_k^{(N+1),u}] = g_k^{(N),u}$$

holds on every k except for a non-frequently visited sequences $k_0, k_1, \ldots$, on which

$$A_{k_l}^{\alpha_{k_l}} g_{k_l}^{(N+1),u} < \max_{u' \in \mathscr{D}_N} [A_k^{u'} g_{k_l}^{(N+1),u}], \quad l = 0, 1, \ldots, \tag{5.28}$$

uniformly, and so,

$$\lim_{n \to \infty} \frac{n}{k_n} = 0. \tag{5.29}$$

Now, we prove that there is another policy in $\bar{\mathscr{D}}_N$, which has a larger $(N+1)$th bias than policy u. Let $\alpha_{k_l}^0$ be the decision rule (with k_l, $l = 0, 1, \ldots$, being the sequence on which (5.28) holds), such that $A_{k_l}^{\alpha_{k_l}^0} g_{k_l}^{(N+1),u} = \max_{u' \in \mathscr{D}_N} [A_k^{u'} g_{k_l}^{(N+1),u}]$. We construct another policy $\tilde{u} = \{\tilde{\alpha}_0, \tilde{\alpha}_1, \ldots\}$ by setting

$$\tilde{\alpha}_k := \begin{cases} \alpha_k, & \forall\, k \neq k_l,\ l = 0, 1, \ldots, \\ \alpha_{k_l}^0, & \forall\, k = k_l,\ l = 0, 1, \ldots. \end{cases} \tag{5.30}$$

Then, $\tilde{u} \in \bar{\mathscr{D}}_N$. Next, setting $u' = \tilde{u}$ in the difference formula (5.20), we get

$$\begin{aligned} & g_0^{(N+1),\tilde{u}}(x) - g_0^{(N+1),u}(x) \\ = & \lim_{K\to\infty} \frac{1}{K} \sum_{k=0}^{K-1} E^{\tilde{u}} \left\{ [A_k^{\tilde{u}} - A_k^u] g_k^{(N+2),u}(X_k^{\tilde{u}}) \Big| X_0^{\tilde{u}} = x \right\} \\ & + \lim_{K\to\infty} \frac{1}{K} \sum_{l=1}^{K-1} (K-l-1) E^{\tilde{u}} \Big\{ [A_{l-1}^{\tilde{u}} - A_{l-1}^u] \\ & \qquad g_{l-1}^{(N+1),u}(X_{l-1}^{\tilde{u}}) \Big| X_0^{\tilde{u}} = x \Big\}. \end{aligned}$$

The first term on its right-hand side is zero because, by construction (5.30), $A_k^{\tilde{u}}$ and A_k^u, $k = 0, 1, \ldots$, differ only at a non-frequently visited sequence satisfying (5.29), which does not change the value of the long-run average. By (5.28) and the construction (5.30), the second term of the right-hand side of the above equation is positive. Thus $g_0^{(N+1),\tilde{u}}(x) > g_0^{(N+1),u}(x)$; and u is not an $(N+1)$th-bias optimal. This means that an $(N+1)$th-bias optimal policy must be in $\bar{\mathscr{D}}_N$. □

Theorem 5.2 (N th-Bias optimality conditions) *Under Assumptions 3.3, 5.7, and 5.8 (for $N+1$ and $N+2$), an Nth-bias optimal policy $u^* \in \mathscr{D}_N$ is $(N+1)$th-bias optimal, if and only if*

(a) $u^* \in \bar{\mathscr{D}}_N$; i.e., its $(N+1)$th biases $g_k^{(N+1),u^*}$, $k = 0, 1, \ldots$, satisfy

$$A_k^{u^*} g_k^{(N+1),u^*} = \max_{u \in \mathscr{D}_N} [A_k^u g_k^{(N+1),u^*}], \quad k = 0, 1, \ldots, \tag{5.31}$$

and

(b) The $(N+2)$th biases $g_k^{(N+2),u^*}$, $k = 0, 1, \ldots$, satisfy

$$A_k^{u^*} g_k^{(N+2),u^*} = \max_{u \in \bar{\mathscr{D}}_N} [A_k^u g_k^{(N+2),u^*}], \quad k = 0, 1, \ldots, \tag{5.32}$$

on every "frequently visited" subsequence of $k = 0, 1, \ldots$; or more precisely, if there exist a subsequence of $k = 0, 1, \ldots$, denoted by $k_0, k_1, \ldots, k_l, \ldots$ and a sequence of decision rules $\alpha_{k_0}^u, \alpha_{k_1}^u, \ldots, \alpha_{k_l}^u, \ldots$, $l = 0, 1, \ldots$, such that (5.32) does not hold on k_l, $l = 0, 1, \ldots$, i.e., $(A_{k_l}^{u^*} g_{k_l}^{(N+2),u^*}) < (A_{k_l}^{\alpha_{k_l}^u} g_{k_l}^{(N+2),u^*})$, then we must have

$$\lim_{n\to\infty} \frac{n}{k_n} = 0. \tag{5.33}$$

Proof By the Nth-bias difference formula (5.20), the proof is similar to those of Theorems 3.2 and 3.5. Condition (5.31) is necessary due to Theorem 5.1. Theorems 5.1 and 5.2 are proved recursively: Using (5.33) in Theorem 5.2, we prove that (5.26), or (5.31), is necessary for $\mathscr{D}_{N+1}$ in Theorem 5.1; then, we prove that (5.33) in condition (b) of Theorem 5.2 holds for $\mathscr{D}_{N+1}$, and so on. The process starts with Lemma 3.12 for $N = 0$. □

5.2.4 The Nth-Bias Optimality Conditions II

The condition (5.26) in Theorem 5.1 and conditions (5.31) and (5.32) in Theorem 5.2 are stated for policies, not actions. However, the number of policies is usually too large, and these conditions are not practically verifiable. For $N > 0$, policies satisfying (5.26) may not form a product space of the action spaces; i.e., there might be correlations among actions at different states in policies in $\bar{\mathscr{D}}_N$. We need to reform the searching spaces.

Choose any policy $u^* \in \bar{\mathscr{D}}_N$ as a reference policy, and define (see (3.46)–(3.48))

$$\mathscr{C}_0 := \{all\ u \in \mathscr{D} :\ A_k^u g_k^{(1),u^*} + f^u = A_k^{u^*} g_k^{(1),u^*} + f^{u^*}, \forall k\}, \tag{5.34}$$

$$\mathscr{C}_N := \{u \in \mathscr{C}_{N-1} : A_k^u g_k^{(N+1),u^*} = A_k^{u^*} g_k^{(N+1),u^*}, \forall k\},\ u^* \in \bar{\mathscr{D}}_N,\ N > 0. \tag{5.35}$$

Componentwise, (5.34) and (5.35) are equivalent to

$$\mathscr{A}_{0,k}(x) = \Big\{\alpha \in \mathscr{A}_k(x) :\ g_k^{(1),u^*}(x) + \eta^* = \sum_{y \in \mathscr{S}_{k+1}} P_k^\alpha(y|x) g_{k+1}^{(1),u^*}(y) + f_k^\alpha(x)\Big\}, \tag{5.36}$$

$$\mathscr{A}_{N,k}(x) = \Big\{\alpha \in \mathscr{A}_{N-1,k}(x) :\ g_k^{(N),u^*}(x) + g_k^{(N+1),u^*}(x) = \sum_{y \in \mathscr{S}_{k+1}} P_k^\alpha(y|x) g_{k+1}^{(N+1),u^*}(y)\Big\},$$
$$x \in \mathscr{S}_k,\ k = 0, 1, \ldots, \quad N = 1, 2, \ldots. \tag{5.37}$$

Then, $\mathscr{C}_N$, $N = 0, 1, \ldots$, can be written in the following form:

$$\mathscr{C}_{N,k} = \prod_{x \in \mathscr{S}_k} \mathscr{A}_{N,k}(x); \tag{5.38}$$

and

$$\mathscr{C}_N = \prod_{k=0}^{\infty} \mathscr{C}_{N,k}. \tag{5.39}$$

We say that u^* is a *reference policy* of $\mathscr{C}_N$, which is a Cartesian product of action spaces. (By Theorem 5.3, $\mathscr{C}_N \subseteq \bar{\mathscr{D}}_N$, so we may choose $u^* \in \mathscr{C}_N \subseteq \mathscr{C}_{N-1}$.)

By Lemma 3.8, we have $\mathscr{C}_0 = \bar{\mathscr{D}}_0$. However, for $N > 0$, this may not be true (the difference formula for the long-run average and those for the Nth biases take different forms). Nevertheless, we have

Theorem 5.3 *Suppose Assumptions 1.1, 1.2, and 5.7, and Assumption 5.8 for N and N + 1 hold. Then, the following statements are true:*

(a) $\mathscr{C}_N \subseteq \bar{\mathscr{D}}_N \subseteq \mathscr{D}_N$.
(b) *Any policy* $u \in \mathscr{C}_N$ *can be chosen as a reference policy in (5.35), resulting in the same* $\mathscr{C}_N$.
(c) *For any two policies* $u, u' \in \mathscr{C}_N$, *it holds that*

$$g_k^{(N+1),u} = g_k^{(N+1),u'} + ce, \ \forall\, k = 0, 1, \ldots, \tag{5.40}$$

with c being a constant.
(d) *If* $\tilde{u} \in \bar{\mathscr{D}}_N - \mathscr{C}_N$, *then there is a policy* $\hat{u} \in \mathscr{C}_N$ *such that* $g_k^{(N+1),\tilde{u}} \le g_k^{(N+1),\hat{u}}$, *for all* $k = 0, 1, \ldots$.

Proof It is shown in Lemma 3.8 that $\mathscr{C}_0 \subseteq \mathscr{D}_0$; in fact, $\mathscr{C}_0 = \mathscr{D}_0$ except for a non-frequently visited sequence of k. Now we prove the lemma for $N \ge 1$.

(a) By (5.35), if $u \in \mathscr{C}_N$, with u^* as a reference point, then $A_k^u g_k^{(N+1),u^*} = A_k^{u^*} g_k^{(N+1),u^*} = g_k^{(N),u^*}$. By definition, $u \in \mathscr{C}_{N-1}$, so by the difference formula (5.20) for u^* and u, we get $g_k^{(N),u^*} = g_k^{(N),u}$, and thus $g_k^{(N),u} = A_k^u g_k^{(N+1),u^*}$. By Lemma 5.2, it holds that $g_k^{(N+1),u} = g_k^{(N+1),u^*} + ce$ for all $k = 0, 1, \ldots$. Thus, by (5.26), we can easily verify that $A_k^u g_k^{(N+1),u} = \max_{u' \in \mathscr{D}_N} [A_k^{u'} g_k^{(N+1),u}] = g_k^{(N),u^*}$, i.e., $u \in \bar{\mathscr{D}}_N$.

(b) Let u be any policy in $\mathscr{C}_N$. From the proof in (a), we have $g_k^{(N+1),u} = g_k^{(N+1),u^*} + ce$ with c being a constant. Let u' be any policy $u' \in \mathscr{D}_N$. Suppose u' is in the $\mathscr{C}_N$ defined by u as the reference policy, i.e., $A_k^{u'} g^{(N+1),u} = A_k^u g^{(N+1),u}$; then, we have

$$\begin{aligned} & A_k^{u'} g^{(N+1),u^*} = A_k^{u'} g^{(N+1),u} \\ = & A_k^u g^{(N+1),u} = A_k^u g^{(N+1),u^*} = A_k^{u^*} g^{(N+1),u^*}. \end{aligned}$$

This means that u' is also in the $\mathscr{C}_N$ defined by reference policy u^*.

(c) This follows from (a) and (b) directly.

(d) Let u^* be the reference policy in (5.35). First, we have[1] $\tilde{u} \in \bar{\mathscr{D}}_N$, $u^* \in \bar{\mathscr{D}}_N \subseteq \mathscr{D}_N$, and the difference formula (by (5.20))

$$\begin{aligned} & g_0^{(N),\tilde{u}}(x) - g_0^{(N),u^*}(x) \\ = & \lim_{K \to \infty} \frac{1}{K} \sum_{k=0}^{K-1} E^{\tilde{u}} \Big\{ [A_k^{\tilde{u}} - A_k^{u^*}] g_k^{(N+1),u^*}(X_k^{u^*}) \Big| X_0^{\tilde{u}} = x \Big\} \\ & + \lim_{K \to \infty} \frac{1}{K} \sum_{l=1}^{K-1} (K - l - 1) E^{\tilde{u}} \Big\{ [A_{l-1}^{\tilde{u}} - A_{l-1}^{u^*}] \\ & \qquad g_{l-1}^{(N),u^*}(X_{l-1}^{\tilde{u}}) \Big| X_0^{\tilde{u}} = x \Big\}. \end{aligned} \tag{5.41}$$

---

[1]The analysis applies to $\tilde{u} \in \mathscr{D}_N$ too.

Because $u^* \in \bar{\mathscr{D}}_N$, every term in the first sum on its right-hand side is non-positive, so the first sum on the right-hand side must be non-positive. Also, by Theorem 5.1, every term in the second sum on its right-hand side is non-positive, so the second sum on the right-hand side must be non-positive too. Therefore, because both $\tilde{u}$ and u^* are Nth-bias optimal policies, the left-hand side of (5.41) is zero, so both sums in (5.41) must be zero. In particular, its first sum is zero:

$$\lim_{K\to\infty} \frac{1}{K} \sum_{k=0}^{K-1} E^{\tilde{u}}\left\{[A_k^{\tilde{u}} - A_k^{u^*}]g_k^{(N+1),u^*}(X_k^{u^*})\middle|X_0^{\tilde{u}} = x\right\} = 0.$$

By (5.26), $A_k^{\tilde{u}} g_k^{(N+1),u^*} \leq A_k^{u^*} g_k^{(N+1),u^*}$, for all $k = 0, 1, \ldots$. Thus, from the above difference formula and with Assumption 5.8, we must have (note that $A_k^{u^*} g_k^{(N+1),u^*} = g_k^{(N),u^*}$, $A_k^{\tilde{u}} g_k^{(N+1),\tilde{u}} = g_k^{(N),\tilde{u}}$, and $g_k^{(N),u^*} = g_k^{(N),\tilde{u}}$)

$$A_k^{\tilde{u}} g_k^{(N+1),u^*} = A_k^{u^*} g_k^{(N+1),u^*} = A_k^{\tilde{u}} g_k^{(N+1),\tilde{u}}$$

for almost all k except, possibly, for a "non-frequently visited" time sequence, denoted as $k_0, k_1, \ldots$, with $\lim_{n\to\infty} \frac{n}{k_n} = 0$, on which $A_{k_l}^{\tilde{u}} g_{k_l}^{(N+1),u^*} < A_{k_l}^{\tilde{u}} g_{k_l}^{(N+1),\tilde{u}}$, uniformly. (Such a sequence may, or may not, exist.)

Now, we construct a policy $\hat{u} \in \mathscr{C}_N$ as follows: For $k \neq k_l, l = 0, 1, \ldots$, set $\hat{u}_k = \tilde{u}_k$; and for $k = k_l, l = 0, 1, \ldots$, we choose any decision rule that satisfies (5.26) (e.g., $\alpha_{k_l}^{u^*}$) and thus $\hat{u} \in \mathscr{C}_N \subseteq \mathscr{C}_{N-1}$. Applying (5.41) to $\tilde{u}$ and $\hat{u}$, we get $g_0^{(N),\hat{u}} = g_0^{(N),\tilde{u}}$. Furthermore, because $u^* \in \mathscr{C}_N$, so by (5.40), we have $g_k^{(N+1),u^*} = g_k^{(N+1),\hat{u}} + ce$, and thus $A_k^{\tilde{u}} g_k^{(N+1),u^*} = A_k^{\tilde{u}} g_k^{(N+1),\hat{u}}$.

Applying Dynkin's formula (3.25) to $A_k^{\tilde{u}}$ and $h(x) = g_k^{(N+1),\hat{u}} - g_k^{(N+1),\tilde{u}}$, we get that, for any $K > k_0$ (note $\tilde{u} \in \bar{\mathscr{D}}_N$),

$$\begin{aligned}
0 \geq & E^{\tilde{u}}\left\{\sum_{k=0}^{K-1} A_k^{\tilde{u}}[g_k^{(N+1),u^*} - g_k^{(N+1),\tilde{u}}](X_k^{\tilde{u}})\middle|X_0^{\tilde{u}} = x\right\} \\
= & E^{\tilde{u}}\left\{\sum_{k=0}^{K-1} A_k^{\tilde{u}}[g_k^{(N+1),\hat{u}} - g_k^{(N+1),\tilde{u}}](X_k^{\tilde{u}})\middle|X_0^{\tilde{u}} = x\right\} \\
= & E^{\tilde{u}}\{[g_K^{(N+1),\hat{u}} - g_K^{(N+1),\tilde{u}}](X_K^{\tilde{u}})|X_0^{\tilde{u}} = x\} \\
& - [g_0^{(N+1),\hat{u}} - g_0^{(N+1),\tilde{u}}](x),
\end{aligned}$$

or

$$\begin{aligned}
& [g_0^{(N+1),\hat{u}} - g_0^{(N+1),\tilde{u}}](x) \\
\geq & E^{\tilde{u}}\{[g_K^{(N+1),\hat{u}} - g_K^{(N+1),\tilde{u}}](X_K^{\tilde{u}})|X_0^{\tilde{u}} = x\}. \qquad (5.42)
\end{aligned}$$

Next, we note that on the non-frequently visited sequence $k_0, k_1, \ldots$, for any $K > 0$, there must be an integer m, such that $k_{m+1} - k_m > K$. Thus,

$$A_k^{\tilde{u}} g_k^{(N+1),\hat{u}} = A_k^{\tilde{u}} g_k^{(N+1),\tilde{u}},$$

for all $k_m + 1 < k < k_{m+1} - 1$. Applying Dynkin's formula (3.25) to this time period, similar to (5.42), we have

$$\begin{aligned} &[g_{k_m+1}^{(N+1),\hat{u}} - g_{k_m+1}^{(N+1),\tilde{u}}](x) \\ =& E^{\tilde{u}}\{[g_{k_{m+1}-1}^{(N+1),\hat{u}} - g_{k_{m+1}-1}^{(N+1),\tilde{u}}](X_{k_{m+1}-1}^{\tilde{u}})|X_{k_m+1}^{\tilde{u}} = x\}. \end{aligned} \tag{5.43}$$

By (5.16), if K is large enough, we have

$$|E^{\tilde{u}}\{[g_{k_{m+1}-1}^{(N+1),\tilde{u}}](X_{k_{m+1}-1}^{\tilde{u}})|X_{k_m+1}^{\tilde{u}} = x\}| < \varepsilon,$$

for any small $\varepsilon > 0$. On the other hand, in the period of $k_m + 1 < k < k_{m+1} - 1$, $\tilde{u}$ and $\hat{u}$ are the same, so for $x \in \mathscr{S}_{k_m+1}$,

$$\begin{aligned} &|E^{\tilde{u}}\{[g_{k_{m+1}-1}^{(N+1),\hat{u}}](X_{k_{m+1}-1}^{\tilde{u}})|X_{k_m+1}^{\hat{u}} = x\}| \\ =& |E^{\hat{u}}\{[g_{k_{m+1}-1}^{(N+1),\hat{u}}](X_{k_{m+1}-1}^{\hat{u}})|X_{k_m+1}^{\hat{u}} = x\}| < \varepsilon. \end{aligned}$$

Therefore, if K is large enough, from the above two inequality and by (5.43), we have

$$|[g_{k_m+1}^{(N+1),\hat{u}} - g_{k_m+1}^{(N+1),\tilde{u}}](x)| < 2\varepsilon,$$

for all $x \in \mathscr{S}_{k_m+1}$. From (5.42), we have (set $K = k_m + 1$)

$$[g_0^{(N+1),\hat{u}} - g_0^{(N+1),\tilde{u}}](x) > -2\varepsilon.$$

Because ε can be any small number, we have

$$g_0^{(N+1),\hat{u}}(x) \geq g_0^{(N+1),\tilde{u}}(x).$$

The theorem is then proved. □

Because the difference formula for average rewards (3.28) takes a simpler form than that for Nth biases (5.19) and (5.20), the analysis for the 1st bias is simpler, as shown in Lemma 3.8.

Note that for any two policies $u, u^* \in \bar{\mathscr{D}}_N$, the equation in (5.35) may not hold at a non-frequently visited sequence of time instants (cf. (5.41)). Therefore, suppose we use any two policies in $\bar{\mathscr{D}}_N$ as the reference policy to construct two $\mathscr{C}_N$'s, respectively; then, these two $\mathscr{C}_N$'s may not be the same. Let u and u^* be any two policies in these

$\mathscr{D}_N$: the space of all the Nth-bias optimal policies

$$\bar{\mathscr{D}}_N := \{u^* \in \mathscr{D}_N : A_k^{u^*} g_k^{(N+1),u^*} = \max_{u \in \mathscr{D}_N} [A_k^u g_k^{(N+1),u^*}], \forall k\}, \quad (5.27)$$

$$\mathscr{C}_N := \{u \in \mathscr{C}_{N-1} : A_k^u g_k^{(N+1),u^*} = A_k^{u^*} g_k^{(N+1),u^*}, \forall k\}, \quad (5.35)$$

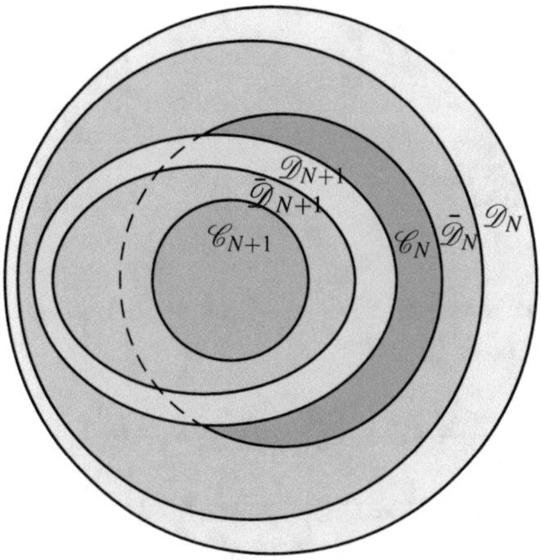

Fig. 5.1 The relations among the policy sets

two $\mathscr{C}_N$'s, respectively. Then, the equation in (5.35) may not hold at a non-frequently visited sequence of time instants.

Figure 5.1 illustrates the relations among the sets $\mathscr{D}_N, \bar{\mathscr{D}}_N, \mathscr{C}_N$, and $\mathscr{D}_{N+1}, \bar{\mathscr{D}}_{N+1}, \mathscr{C}_{N+1}$. Corresponding policies in $\mathscr{D}_N$ and $\bar{\mathscr{D}}_N$, and in $\bar{\mathscr{D}}_N$ and $\mathscr{C}_N$, respectively, differ only at a non-frequently visited sequence of time instants.

Theorem 5.3d shows that to search for an $(N + 1)$th-bias optimal policy, we need only to look into $\mathscr{C}_N$. Therefore, the $(N + 1)$th-bias optimal policy defined in (5.18) is now

$$g_k^{(N+1),u^{(N+1),*}} = \max_{u \in \mathscr{C}_N}\{g_k^{(N+1),u}\} =: g_k^{(N+1),*}, \quad (5.44)$$

for all $k = 0, 1, \ldots$. By Theorem 5.3c, this in fact maximizes the constant c in (5.40). Similar results for time-homogeneous MDPs can be found in Cao (2007).

By the structure of (5.34)–(5.39), the space $\mathscr{C}_N, N = 0, 1, \ldots$, is the Cartesian product of the action spaces; this allows us to search for an Nth-bias optimal policy (5.44) in the action spaces, not in the policy space. This is extremely important because the dimension of a policy space can be huge. On the other hand, $\bar{\mathscr{D}}_N$ and $\mathscr{D}_N$ may not take the product form.

With (5.44), we now turn to the optimality conditions. First, we have

Lemma 5.3 *Let $u^* = (\mathbb{P}^*, \boldsymbol{f}^*) \in \bar{\mathscr{D}}_N$ be an $(N + 1)$th-bias optimal policy, with $A_k^{u^*}$ and $g_k^{(N+2),u^*}$, $k = 0, 1, \ldots, N \geq 1$, and assume that Assumptions 1.1, 1.2, and 5.7 hold. Any other Nth-bias optimal policy $u = (\mathbb{P}, \boldsymbol{f}) \in \mathscr{C}_N$ (with u^* as a reference policy) is $(N + 1)$th-bias optimal, i.e., $u \in \mathscr{D}_{N+1}$, if*

$$A_k^u g_k^{(N+2),u^*} = A_k^{u^*} g_k^{(N+2),u^*}, \quad k = 0, 1, \ldots. \tag{5.45}$$

Componentwise, this is

$$\sum_{y \in \mathscr{S}_{k+1}} P_k^{\alpha_k(x)}(y|x) g_{k+1}^{(N+2),u^*}(y) = \sum_{y \in \mathscr{S}_{k+1}} P_k^{\alpha_k^*(x)}(y|x) g_{k+1}^{(N+2),u^*}(y),$$

$x \in \mathscr{S}_k$, $k = 0, 1, \ldots$, where $u = \{\alpha_k(x), x \in \mathscr{S}_k, k = 0, 1, \ldots\}$, and $u^* = \{\alpha_k^*(x), x \in \mathscr{S}_k, k = 0, 1, \ldots\}$. For $N = 0$, it is

$$P_k^u g_{k+1}^{(1),u^*} + f_k^u = P_k^{u^*} g_{k+1}^{(1),u} + f_k^{u^*}, \quad k = 0, 1, \ldots.$$

Proof First, consider the case with $N = 0$. Suppose u^* is 1st-bias optimal, and let $u \in \mathscr{C}_0$ be a long-run average optimal policy. Set $u = u^*$ and $u' = u$ in the difference formula (5.19). Because $u \in \mathscr{C}_0$, we have $(P_k^u g_{k+1}^{(1),u^*} + f_k^u)(x) = (P_k^{u^*} g_{k+1}^{(1),u} + f_k^{u^*})(x)$, $x \in \mathscr{S}_k$, $k = 0, 1, \ldots$. Thus, the second term on right-hand side of (5.19) is zero, and it becomes

$$\begin{aligned} & g_0^{(1),u}(i) - g_0^{(1),u^*}(i) \\ = & \lim_{K \to \infty} \frac{1}{K} \sum_{k=0}^{K-1} E^u \left\{ [A_k^u - A_k^{u^*}] g_k^{(2),u^*}(X_k^u) \middle| X_0^u = i \right\}. \end{aligned}$$

Therefore, if (5.45) with $N = 0$, i.e., $A_k^{u^*} g_k^{(2),u^*} = A_k^u g_k^{(2),u^*}$, holds for all k, then $g_0^{(1),u}(i) = g_0^{(1),u^*}(i)$, $i \in \mathscr{S}_0$, and u is 1st-bias optimal.

Next, for $N \geq 1$, because $u \in \mathscr{C}_N$, we have $A_k^{u^*} g_k^{(N+1),u^*} = A_k^u g_k^{(N+1),u^*}$ all k. Thus, the $(N+1)$th-bias difference formula (5.20) becomes

$$\begin{aligned} & g_0^{(N+1),u}(x) - g_0^{(N+1),u^*}(x) \\ = & \lim_{K \to \infty} \frac{1}{K} \sum_{k=0}^{K-1} E^u \left\{ [A_k^u - A_k^{u^*}] g_k^{(N+2),u^*}(X_k^u) \middle| X_0^u = x \right\}. \end{aligned} \tag{5.46}$$

Thus, if (5.45) holds, then $g_0^{(N+1),u}(x) = g_0^{(N+1),u^*}(x)$, i.e., u is another $(N+1)$th-bias optimal policy. □

Theorem 5.4 (Nth-Bias optimality conditions) *Under Assumptions 1.1, 1.2, 5.7, and 5.8, an Nth-bias optimal policy $u^* \in \bar{\mathscr{D}}_N$ is $(N+1)$th-bias optimal, if its $(N+2)$th bias $g_k^{(N+2),u^*}$, $k = 0, 1, \ldots$, satisfy*

$$(A_k^{u^*} g_k^{(N+2),u^*})(x) = \max_{u \in \mathscr{C}_N} \{(A_k^u g_k^{(N+2),u^*})(x)\}, \tag{5.47}$$

$x \in \mathscr{S}_k$, for all $k = 0, 1, \ldots$, where $\mathscr{C}_N$ is the set defined in (5.35) and (5.39) with u^* as a reference policy. Componentwise, it is

$$\sum_{y\in\mathscr{S}_{k+1}} P_k^{\alpha_k^*(x)}(y|x)g_{k+1}^{(N+2),u^*}(y) = \max_{\alpha\in\mathscr{A}_{N,k}(x)}\Big\{\sum_{y\in\mathscr{S}_{k+1}} P_k^{\alpha}(y|x)g_{k+1}^{(N+2),u^*}(y)\Big\},$$

$x \in \mathscr{S}_k$, and $\mathscr{A}_{N,k}(x)$ is defined in (5.36) and (5.37), $k = 0, 1, \ldots$.

Proof The theorem follows directly from (5.26) and the Nth-bias difference formula (5.46). □

Remark 1 (a) Because of the under-selectivity due to the difference formula (5.46), (5.47) does not need to hold for every k; i.e., it may not need to hold in any non-frequently visited time instants.

(b) The logic behind the theorem is as follows. First, suppose that we have an Nth-bias optimal policy $u^* \in \mathscr{D}_N$. Then, by induction,

$$(A_k^{u^*} g_k^{(N+1),u^*})(x) = \max_{u\in\mathscr{C}_N}\{(A_k^u g_k^{(N+1),u^*})(x)\}$$

holds except that it may not hold at a non-frequently visited sequence of time instants for some policies. However, we choose one in $\bar{\mathscr{D}}_N$; i.e., $u^* \in \bar{\mathscr{D}}_N$. This is natural and required by Theorem 5.1.

(c) The optimality conditions can be checked in the action spaces, not in the policy space. □

Now, based on Theorem 5.4, we define a system of equations:

$$g_k^{(1),u^*} + \eta^{u^*} e = \max_{u\in\mathscr{D}}\{P_k^u g_{k+1}^{(1),u^*} + f_k^u\}, \tag{5.48}$$

$$g_k^{(2),u^*} + g_k^{(1),u^*} = \max_{u\in\mathscr{C}_0}\{P_k^u g_{k+1}^{(2),u^*}\}, \tag{5.49}$$

$$g_k^{(3),u^*} + g_k^{(2),u^*} = \max_{u\in\mathscr{C}_1}\{P_k^u g_{k+1}^{(3),u^*}\}, \tag{5.50}$$

$$\cdots\cdots\cdots$$

$$g_k^{(N+1),u^*} + g_k^{(N),u^*} = \max_{u\in\mathscr{C}_{N-1}}\{P_k^u g_{k+1}^{(N+1),u^*}\}, \tag{5.51}$$

where $\mathscr{C}_k, k = 0, 1, \ldots, N-1$, are the sets defined in (5.35) with u^* as the reference policy. By Theorem 5.4, a policy $u^* \in \mathscr{D}$ satisfying Eqs. (5.48)–(5.51) is an Nth-bias optimal policy.

All the optimality equations can be verified componentwise. A policy $u^* \in \mathscr{D}$, $u^* := \{\alpha_k^*(x), x \in \mathscr{S}_k, k = 0, 1, \ldots\}$, is an Nth-bias optimal, if

$$\sum_{y \in \mathscr{S}_{k+1}} P_k^{\alpha_k^*(x)}(y|x) g_{k+1}^{(1),u^*}(y) + f_k^{\alpha_k^*(x)}(x)$$
$$= \max_{\alpha \in \mathscr{A}_k(x)} \Big\{ \sum_{y \in \mathscr{S}_{k+1}} P_k^{\alpha}(y|x) g_{k+1}^{(1),u^*}(y) + f_k^{\alpha}(x) \Big\}, \; x \in \mathscr{S}_k,$$
$$\sum_{y \in \mathscr{S}_{k+1}} P_k^{\alpha_k^*(x)}(y|x) g_{k+1}^{(m+1),u^*}(y)$$
$$= \max_{\alpha \in \mathscr{A}_{m-1,k}(x)} \Big\{ \sum_{y \in \mathscr{S}_{k+1}} P_k^{\alpha}(y|x) g_{k+1}^{(m+1),u^*}(y) \Big\},$$
$$x \in \mathscr{S}_k, \; 1 \le m \le N,$$

where $\mathscr{A}_k(x)$ is the action space at x and k, and $\mathscr{A}_{m,k}(x)$, $m = 0, 1, \ldots$, are defined in (5.37).

Let u_{all} be a policy that is Nth-bias optimal for all $N = 0, 1, \ldots$; i.e., $u_{all} \in \cap_{N \ge 0} \mathscr{D}_N$. For discrete policy spaces, because $\mathscr{D}_{N+1} \subseteq \mathscr{D}_N$ and $\mathscr{D}_N \neq \emptyset$ for all N, we have $\cap_{N \ge 0} \mathscr{D}_N \neq \emptyset$, and such a u_{all} does exist. A policy u_{all} takes care of all the Nth biases, reflecting the transient behavior at all levels, and no under-selectivity issue exists. In the next section, we prove that u_{all} is in fact a Blackwell optimal policy, and the Nth-bias optimality is equivalent to the sensitive discount optimality.

5.3 Laurent Series, Blackwell Optimality, and Sensitivity Discount Optimality

5.3.1 The Laurent Series

Consider a TNHMC with state spaces $\mathscr{S}_k$, and transition probability matrices P_k and reward functions f_k, $k = 0, 1, \ldots$; the *discounted reward* is defined by (cf. (4.72)):

$$\upsilon_\beta(x) := E\left\{ \sum_{l=0}^{\infty} \beta^l f_l(X_l) \Big| X_0 = x \right\}, \quad x \in \mathscr{S}_0, \;\; 0 < \beta < 1.$$

Denote $\upsilon_\beta = (\upsilon_\beta(1), \ldots, \upsilon_\beta(S_0))^T$. Set $\beta = (1+\rho)^{-1}$, or $\rho = (1-\beta)/\beta$. $0 < \beta < 1$ implies $\infty > \rho > 0$.

Lemma 5.4 *The following Laurent series expansion in $g_0^{(N)}$, $g_0^{(0)} := \eta e$, holds,*

$$\upsilon_\beta = (1+\rho) \sum_{N=-1}^{\infty} \rho^N g_0^{(N+1)}, \quad \rho > 0, \tag{5.52}$$

Proof First, by (5.17), we have that, for any $x \in \mathscr{S}_0$,

$$\sum_{N=0}^{\infty} \rho^N g_0^{(N+1)}(x)$$
$$= \sum_{k=0}^{\infty} \Big\{ \Big[\sum_{N=0}^{\infty} (-\rho)^N \binom{N+k}{N} \Big] E\{[f_k(X_k) - \eta] | X_0 = x\} \Big\}$$
$$= \sum_{k=0}^{\infty} \Big\{ \beta^{k+1} E\{[f_k(X_k) - \eta] | X_0 = x\} \Big\},$$

in which the binomial series $\frac{1}{(1-z)^{k+1}} = \sum_{N=0}^{\infty} \binom{N+k}{N} z^N, \quad |z| < 1$, is used. Therefore,

$$(1+\rho) \sum_{N=1}^{\infty} \rho^{N-1} g_0^{(N)}(x)$$
$$= \sum_{k=0}^{\infty} \Big\{ \beta^k E\{[f_k(X_k) - \eta] | X_0 = x\} \Big\}$$
$$= \sum_{k=0}^{\infty} \Big\{ \beta^k E[f_k(X_k) | X_0 = x] \Big\} - (\sum_{k=0}^{\infty} \beta^k) \eta$$
$$= \sum_{k=0}^{\infty} \Big\{ \beta^k E[f_k(X_k) | X_0 = x] \Big\} - \frac{1+\rho}{\rho} \eta.$$

On the other hand, the first term in (5.52) is

$$(1+\rho)\rho^{-1} g_0^{(0)}(x) = \frac{1+\rho}{\rho} \eta.$$

Finally, adding the above two equations together, we obtain the Laurent series expansion in $g_0^{(N)}(x)$, (5.52). □

Similar expansions in $g_k^{(N+1)}(x)$'s hold at all the other time instants $k > 0$. For any $k = 0, 1, \ldots$, the discounted performance is defined by

$$\upsilon_{\beta,k}(x) := E\left\{ \sum_{l=0}^{\infty} \beta^l f_{k+l}(X_{k+l}) \Big| X_k = x \right\}, \quad x \in \mathscr{S}_k, \quad 0 < \beta < 1;$$

with $\upsilon_\beta(x) \equiv \upsilon_{\beta,0}(x)$, $x \in \mathscr{S}_0$. The Laurent series expansion at k is

$$\upsilon_{\beta,k}(x) = (1+\rho) \sum_{N=-1}^{\infty} \rho^N g_k^{(N+1)}(x), \quad x \in \mathscr{S}_k, \ \rho > 0.$$

The Laurent series for THMCs is presented in, e.g., Cao (2007), Cao and Zhang (2008b), Guo and Hernández-Lerma (2009), Hernández-Lerma and Lasserre (1996), Puterman (1994).

5.3.2 *Blackwell Optimality*

A policy $u_b \in \mathscr{D}$ is said to be *Blackwell* optimal, if for any policy $u \in \mathscr{D}$, there exists a β_u (depending on u), $0 \le \beta_u < 1$, such that

$$\upsilon_\beta^{u_b} \ge \upsilon_\beta^u, \quad \text{for all } \beta \in [\beta_u, 1). \tag{5.53}$$

Set $\rho^* = \inf_{u \in \mathscr{D}} \rho_u$. If $\rho^* > 0$, then

$$\upsilon_\beta^{u^*} \ge \upsilon_\beta^u, \quad \forall u \in \mathscr{D}, \ 0 \le \rho < \rho^*.$$

Such a u^* is called a *strong Blackwell optimal policy*. For time-homogeneous MDPs with a finite number of policies in $\mathscr{D}$, Blackwell optimality is the same as strong Blackwell optimality.

Theorem 5.5 *A policy u^* is a Blackwell optimal policy if and only if it is an Nth-bias optimal for all $N \ge 0$; i.e., $u_b = u_{all}$.*

Proof (a) The "If" part: Since $\rho \to 0$ when $\beta \to 1$, the definition of Blackwell optimality is equivalent to the following: A policy $u_b \in \mathscr{D}$ is a Blackwell optimal policy if for any policy $u \in \mathscr{D}$, there exists a ρ_u for which $\upsilon_\beta^{u_b} \ge \upsilon_\beta^u$ for all $0 < \rho \le \rho_u$.

Suppose u^* maximizes all $g_0^{(N),u}$, $N \ge 0$. Let $u \in \mathscr{D}$ be any policy. Suppose that for some $N = 0, 1, \ldots, u \notin \mathscr{D}_N$. Then, we set N' to be the minimum of such N's: $N' := \min\{N \ge 0 : u \notin \mathscr{D}_N\}$. For u^* and u, by the Laurant Expansion (5.52), we have

$$\begin{aligned} &\upsilon_\beta^{u^*}(x) - \upsilon_\beta^u(x) \\ =&(1+\rho) \sum_{N=-1}^{\infty} \rho^N [g_0^{(N+1),u^*}(x) - g_0^{(N+1),u}(x)], \quad \rho > 0. \end{aligned} \tag{5.54}$$

First, we consider the case $N' < \infty$. Then, by definition of N', for all $k < N'$, we have $g_0^{(k),u^*}(x) - g_0^{(k),u}(x) = 0$; the above equation becomes

$$\begin{aligned} &\upsilon_\beta^{u^*}(x) - \upsilon_\beta^u(x) \\ =&(1+\rho) \sum_{k=N'-1}^{\infty} \rho^k [g_0^{(k+1),u^*}(x) - g_0^{(k+1),u}(x)], \quad \rho > 0. \end{aligned}$$

Next, we left-multiply both sides of the above equation with $(1+\rho)^{-1}\rho^{-(N'-1)}$ and obtain

$$\begin{aligned}
&(1+\rho)^{-1}\rho^{-(N'-1)}[\upsilon_\beta^{u^*}(x)-\upsilon_\beta^{u}(x)]\\
&=[g_0^{(N'),u^*}(x)-g_0^{(N'),u}(x)]\\
&\quad+\sum_{k=1}^{\infty}\rho^k[g_0^{(N'+k),u^*}(x)-g_0^{(N'+k),u}(x)],\quad \rho>0.
\end{aligned} \tag{5.55}$$

Since $u\notin\mathscr{D}_{N'}$, so $g_0^{(N'),u^*}(x)-g_0^{(N'),u}(x)>0$, $x\in\mathscr{S}_0$. From (5.55), we can find a ρ_u such that

$$(1+\rho)^{-1}\rho^{-(N'-1)}[\upsilon_\beta^{u^*}(x)-\upsilon_\beta^{u}(x)]>0,\quad 0<\rho<\rho_u;$$

i.e.,

$$\upsilon_\beta^{u^*}(x)-\upsilon_\beta^{u}(x)>0,\quad 0<\rho<\rho_u. \tag{5.56}$$

Next, we assume $N'=\infty$; i.e., $u\in\mathscr{D}_N$ for all $N\geq 0$, this means $g_0^{(N),u^*}(x)-g_0^{(N),u}(x)=0$, $x\in\mathscr{S}_0$, for all $N\geq 0$. By (5.54), we have $\upsilon_\beta^{u}(x)=\upsilon_\beta^{u^*}(x)$, $x\in\mathscr{S}_0$, $0<\beta<1$. This equation and (5.56) imply that u^* is Blackwell optimal in (5.53). In addition, if $N'=\infty$, then u is a Blackwell policy as well.

(b) The "Only if" part. Let u^* be a Blackwell optimal policy, and $u\in\mathscr{D}$ be any policy. We have $g_0^{(0),u}=\eta^u$; and (5.54) is

$$\begin{aligned}
&\upsilon_\beta^{u^*}(x)-\upsilon_\beta^{u}(x)\\
&=\frac{1+\rho}{\rho}[\eta^{u^*}-\eta^u]+(1+\rho)[g_0^{(1),u^*}(x)-g_0^{(1),u}(x)]\\
&\quad+(1+\rho)\sum_{N=1}^{\infty}\rho^N[g_0^{(N+1),u^*}(x)-g_0^{(N+1),u}(x)].
\end{aligned} \tag{5.57}$$

It is clear that when $\rho\to 0$, $\eta^{u^*}-\eta^u$ is the major term on the right-hand side; and since $\upsilon_\beta^{u^*}(x)-\upsilon_\beta^{u}(x)\geq 0$, we must have $\eta^{u^*}-\eta^u\geq 0$ for any $u\in\mathscr{D}$; i.e., $u^*\in\mathscr{D}_0$ is a long-run average optimal policy.

Next, for any long-run average optimal policies $u\in\mathscr{D}_0$, we have $\eta^u=\eta^{u^*}$, and the first term in (5.57) disappears, and as $\rho\to 0$, $g_0^{(1),u^*}(x)-g_0^{(1),u}(x)$ becomes the major term in the difference $\upsilon_\beta^{u^*}(x)-\upsilon_\beta^{u}(x)(\geq 0)$, and thus we must have $g_0^{(1),u^*}(x)\geq g_0^{(1),u}(x)$, $u\in\mathscr{D}_0$. That is, $u^*\in\mathscr{D}_1$ is a bias optimal policy.

The rest can be proved by induction. Suppose we have proved that u^* is Nth-bias optimal. Let $u\in\mathscr{D}_N$ be any Nth-bias optimal policy. Then, the first $N+1$ terms in (5.57) are all zero, and it becomes

$$\upsilon_\beta^{u^*}(x) - \upsilon_\beta^u(x) = (1+\rho)\sum_{k=N}^{\infty}\rho^k[g_0^{(k+1),u^*}(x) - g_0^{(k+1),u}(x)].$$

Multiplying both sides with $(1+\rho)^{-1}\rho^{-N}$, we have

$$\begin{aligned}&(1+\rho)^{-1}\rho^{-N}[\upsilon_\beta^{u^*}(x) - \upsilon_\beta^u(x)]\\ =&[g_0^{(N+1),u^*}(x) - g_0^{(N+1),u}(x)]\\ &+\sum_{k=1}^{\infty}\rho^k[g_0^{(N+k+1),u^*}(x) - g_0^{(N+k+1),u}(x)].\end{aligned} \tag{5.58}$$

So as $\rho \to 0$, $[g_0^{(N+1),u^*}(x) - g_0^{(N+1),u}(x)]$ is the major term on the right-hand side, and other terms go to zero. Since $(1+\rho)^{-1}\rho^{-N}[\upsilon_\beta^{u^*}(x) - \upsilon_\beta^u(x)] \geq 0$, for all $0 \leq \rho < \rho_u$, we conclude that $g_0^{(N+1),u^*}(x) \geq g_0^{(N+1),u}(x)$, $u \in \mathscr{D}_N$; i.e., $u^* \in \mathscr{D}_{N+1}$ is $(N+1)$th-bias optimal, for all $N \geq 1$. □

In (5.58), we let $\rho \downarrow 0$ and obtain that for $u \in \mathscr{D}_N$,

$$\begin{aligned}&\lim_{\rho\downarrow 0}\rho^{-N}[\upsilon_\beta^{u^*}(x) - \upsilon_\beta^u(x)]\\ =&\lim_{\rho\downarrow 0}(1+\rho)^{-1}\rho^{-N}[\upsilon_\beta^{u^*}(x) - \upsilon_\beta^u(x)]\\ =&g_0^{(N+1),u^*}(x) - g_0^{(N+1),u}(x), \quad x \in \mathscr{S}_0.\end{aligned}$$

On the other hand, if $\lim_{\rho\downarrow 0}\rho^{-N}[\upsilon_\beta^{u^*}(x) - \upsilon_\beta^u(x)] = 0$, we must have $\lim_{\rho\downarrow 0}\rho^{-k}[\upsilon_\beta^{u^*}(x) - \upsilon_\beta^u(x)] = 0$ for all $-1 \leq k < N$.

5.3.3 *Sensitivity Discount Optimality*

Denote the optimal discount value as

$$\upsilon_\beta^*(x) = \max_{u\in\mathscr{D}}\{\upsilon_\beta^u(x)\}, \quad x \in \mathscr{S}_0,\ 0 < \beta < 1.$$

Again, we assume that the maximum is reachable (under some conditions such as compactness of $\mathscr{D}$). A policy $u^* \in \mathscr{D}$ is said to be *N-discount optimal*, if

$$\limsup_{\rho\downarrow 0}\{\rho^{-N}[\upsilon_\beta^{u^*}(x) - \upsilon_\beta^u(x)]\} \geq 0, \quad x \in \mathscr{S}_0, \tag{5.59}$$

for all $u \in \mathscr{D}$. A policy $u^* \in \mathscr{D}$ is said to be *strong N-discount optimal*, if

$$\lim_{\rho \downarrow 0}\{\rho^{-N}[\upsilon_\beta^{u^*}(x) - \upsilon_\beta^*(x)]\} = 0, \quad x \in \mathscr{S}_0.$$

Evidently, strong N-discount optimality implies N-discount optimality, and strong N-discount optimality implies strong k-discount optimality, for all $-1 \le k \le N$.

Theorem 5.6 *A strong N-discount optimal policy is an* $(N+1)$*th-bias optimal policy, and vice versa.*

Proof (a) "⇒": By Theorem 5.5, a Blackwell optimal policy u_b has optimal Nth biases for all $N \ge 0$, and denote them as $g_0^{(N),*}(x)$, $x \in \mathscr{S}_0$, By the Laurent series (5.52) for u_b and any policy u, we have

$$\begin{aligned} &\upsilon_\beta^u(x) - \upsilon_\beta^{u_b}(x) \\ =&\frac{1+\rho}{\rho}[\eta^u - \eta^*] + (1+\rho)[g_0^{(1),u}(x) - g_0^{(1),*}(x)] \\ &+ (1+\rho)\sum_{N=1}^{\infty} \rho^N [g_0^{(N+1),u}(x) - g_0^{(N+1),*}(x)]. \end{aligned} \tag{5.60}$$

For $N = -1$, $\rho^{-N} = \rho$, and by (5.60), we have

$$\lim_{\rho \downarrow 0}\{\rho[\upsilon_\beta^u(x) - \upsilon_\beta^{u_b}(x)]\} = \eta^u - \eta^*.$$

Thus, if u is strong $(N=-1)$-discount optimal, then by (5.59), $\lim_{\rho\to 0}\{\rho[\upsilon_\beta^u(x) - \upsilon_\beta^{u_b}(x)]\} \ge 0$; which implies $\eta^u \ge \eta^*$. So $\eta^u = \eta^*$ because η^* is optimal, and u is a 0th bias, or the long-run average optimal policy.

For $N = 0$, if $\eta^u < \eta^*$, then by the first term on the right-hand side of (5.60), $\lim_{\rho\downarrow 0}[\upsilon_\beta^u(x) - \upsilon_\beta^{u_b}(x)] = -\infty$, and u cannot be $(N=0)$-discount optimal. Thus, if u is 0-discount optimal, then we must have $\eta^u = \eta^*$, and u must be a long-run average optimal policy. In addition, by (5.60) again,

$$\lim_{\rho \downarrow 0}[\upsilon_\beta^u(x) - \upsilon_\beta^*(x)] = g_0^{(1),u}(x) - g_0^{(1),*}(x).$$

Therefore, the strong 0-discount optimality, $\lim_{\rho\to 0}[\upsilon_\beta^u(x) - \upsilon_\beta^*(x)] \ge 0$, implies $g_0^{(1),u}(x) \ge g_0^{(1),*}(x), x \in \mathscr{S}_0$; thus, $g_0^{(1),u}(x) = g_0^{(1),*}(x)$, and u is (first) bias optimal.

Next, if u is strong 1-discount optimal, then it is also strong 0-discount optimal, so it is long-run average and (first) bias optimal. Then, the first two terms on the right-hand side of (5.60) are zero, and this leads to

$$\lim_{\rho \downarrow 0}\rho^{-1}[\upsilon_\beta^u(x) - \upsilon_\beta^*(x)] = g_0^{(2),u}(x) - g_0^{(2),*}(x).$$

Thus, the 1-discount optimality, $\lim_{\rho\to 0}\rho^{-1}[\upsilon^u_\beta(x)-\upsilon^*_\beta(x)]\geq 0$, implies $g_0^{(2),u}(x)\geq g_0^{(2),*}(x)$, $x\in\mathscr{S}_0$, i.e., u is 2nd-bias optimal.

Similarly, we may prove that strong N-discount optimality implies $(N+1)$th-bias optimal for all $N\geq -1$.

(b) "$\Leftarrow$": This can be proved by reversing the above arguments. Similar to (5.60), for any two policies u and u^*, we have

$$\begin{aligned}
&\upsilon^{u^*}_\beta(x)-\upsilon^u_\beta(x)\\
=&\frac{1+\rho}{\rho}[\eta^{u^*}-\eta^u]+(1+\rho)[g_0^{(1),u^*}(x)-g_0^{(1),u}(x)]\\
&+(1+\rho)\sum_{N=1}^{\infty}\rho^N[g_0^{(N+1),u^*}(x)-g_0^{(N+1),u}(x)].
\end{aligned}\tag{5.61}$$

If u^* is long-run average (0th) optimal, then $\eta^{u^*}\geq\eta^u$, and $\lim_{\rho\downarrow 0}\{\rho[\upsilon^{u^*}_\beta(x)-\upsilon^u_\beta(x)]\}=\eta^{u^*}-\eta^u\geq 0$, for any u. Thus, u^* is strong (-1)-discount optimal. Next, if u^* is (1th) bias optimal, then $g_0^{(1),u^*}\geq g_0^{(1),u}$ for all u; and it is also long-run average optimal and $\eta^{u^*}\geq\eta^u$. By (5.61), we have $\lim_{\rho\downarrow 0}\{[\upsilon^{u^*}_\beta(x)-\upsilon^u_\beta(x)]\}\geq 0$, for all u, and u^* is strong 0-discount optimal. The case for $N\geq 1$ can be proved similarly. □

The problem of sensitive discount optimality for THMCs was first proposed in Veinott (1969), and further discussed in Puterman (1994), Guo and Hernández-Lerma (2009), Jasso-Fuentes and Hernández-Lerma (2009a), Zhang and Cao (2009), etc. The goal is to overcome the under-selectivity issue of the long-run average. In the approach, the Blackwell optimality is approximated by the N-discount optimality, with the discount factor β approaching one. Our approach shows that the same goal can be achieved by the Nth-bias optimality without discounting. This approach was proposed in Cao and Zhang (2008b), Cao (2007) for THMCs. Here we extend it to TNHMCs.

5.3.4 On Geometrical Ergodicity

The notion of "Transient" is relative to that of "steady-state". Thus, to discuss the transient performance, we need to assume that the steady-state long-run average (5.2), in the form of limit, exists. To illustrate the ideas, let us assume that all $\mathscr{S}_k$, $k=0,1,\ldots$, have the same dimension S. If the steady state is reachable; then, $\lim_{k\to\infty}E[f_k(X_k)|X_0=x]=\eta$ for all $x\in\mathscr{S}_0$, or equivalently,

$$\lim_{k\to\infty}[P_0P_1\ldots P_k]f_k=\eta e.\tag{5.62}$$

This equation puts restrictions on P_k and f_k. A sufficient condition for (5.62) is that the Markov chain is (strongly) ergodic; i.e., $\lim_{k\to\infty}[P_0 P_1 \dots P_k] = e\pi$ for a stationary distribution $\pi := (\pi(1), \dots, \pi(S))$, and $\pi f_k = \eta$, independent of k. It implies that if k is large enough, we have $(e\pi)P_k \approx e\pi$, or $\pi P_k \approx \pi$; i.e., as $k \to \infty$, P_k, has the same stationary probability.

Another "weaker" condition is that the sequence $P_1, P_2, \dots, P_k, \dots$ is weakly ergodic (cf. (2.4)), i.e., $\lim_{k\to\infty}[P_1 P_2 \dots P_k - e\pi_k] = 0$, where π_k may be different for different k. In addition, for (5.62) to hold, it also requires $\lim_{k\to\infty} \pi_k f_k = \eta$.

To prove that the Nth biases exist, it requires that the long-run average (5.2) converges fast enough. Thus, we need the geometrical ergodicity defined in Assumption 5.7. The research on ergodicity (weak or strong) of nonhomogeneous Markov chains itself is an active area since 1950s (Anthonisse and Tijms 1977; Chan 1989; Chevalier et al. 2017; Daubechies and Lagarias 1992; Hajnal 1958; Mukhamedov 2013; Zeifman 1994), so we refer to these works for specific conditions for geometrical ergodicity.

The requirement of geometrical ergodicity is not as restrictive as it looks. It is rather a description of the scope of the problem: the transient behavior only makes sense when the system can reach its steady state. Many real-world physical, engineering, economic, and social systems reach the steady state after an initial period.

Next, we give some simple examples to illustrate the possible application. First, we note that Assumption 5.7 can be modified as follows.

Assumption 5.9 There exist a $1 > q > 0$, a constant C, and an integer $K > 0$, such that $E\{|f_{l+k}(X_{l+k}) - \eta| \,\big|X_k = x\} < Cq^l$, $x \in \mathscr{S}_k$, for all $k \geq K$ and $l > 0$.

In fact, Assumption 5.9 implies Assumption 5.7; but sometimes the former is easier to verify.

Example 5.1 Let $P_0, P_1, \dots, P_k$ be a set of finite ergodic stochastic matrices, and P_k is a square matrix. Define a policy as

$$\mathbb{P} := \{P_0, P_1, \dots, P_k, P_k, \dots, P_k, \dots\}. \tag{5.63}$$

Assumption 5.9 holds. All the Nth biases for $\mathbb{P}$ exist.

Many engineering, physical, and financial systems may reach the steady state after an initial time. For examples, when an electrical circuit is turned on, it will reach a steady state after oscillating for a while. In the wake of a financial crisis, a stock market will reach the steady state after an initial transition period. In launching a satellite, it will go around the earth a few times before reaching its final orbit. All these systems can be modeled approximately by the TNHMCs in the form of (5.63), and their transient performance can be optimized using the theory of the Nth-bias optimization developed in this chapter. □

If the limit in (1.7) does not exist, then we need to use the "lim inf" (1.6) to define the long-run average. The question is, how to define the transient performance?

5.4 Discussions and Extensions

In this chapter, we have developed the Nth-bias optimality theory for TNHMCs. The Nth-bias optimality equations were derived. The Nth biases, $N = 0, 1, \ldots$, measure the transient performance of a TNHMC at different levels. The optimal policy for all the Nth biases, $N = 0, 1, \ldots$, overcomes the under-selectivity issue. In addition, we have shown that such an "all-bias" optimal policy is a Blackwell optimal policy, and an Nth-bias optimal policy is an $(N - 1)$-discount optimal policy, and vice versa.

Because of the non-stationary nature, a TNHMC is in the transient status at all times, and some optimality conditions may not need to hold for any finite period. Properties such as ergodicity cannot be expected; and we have to use the notions of confluencity and weak ergodicity instead. The optimization approach used is the relative optimization, which is based on a direct comparison of the performance measures, or the biases, of any two policies. The results of this chapter illustrate the advantages of this approach.

Finally, to avoid obscuring the main principles with technicalities, we have made various assumptions, e.g., the finite states and boundness of rewards. To relax these assumptions will be further research topics. Also, in this book, we have studied only Markov policies and also have assumed that an optimal policy exists. Thus, there are questions about whether an optimal Markov policy does exist, and whether a history-dependent policy might perform better. Some works in this direction are ongoing (Cao 2020; Xia et al. 2020).

Glossary

α_k	Decision rule at time k; $\alpha = \alpha_k(x)$: action taken at state x and time k
A_k	Infinitesimal generator at time k, for any sequence of functions $h_k(x)$, $x \in \mathscr{S}_k$, $k = 0, 1, \ldots$, $A_k h_k(x) := \sum_{y \in \mathscr{S}_{k+1}} P_k(y\|x) h_{k+1}(y) - h_k(x)$
$\mathscr{A}_k$	Space of all decision rules at time k, $\mathscr{A}_k(x) = \prod_{x \in \mathscr{S}_k} \mathscr{A}_k(x)$
$\mathscr{A}_k(x)$	Space of all actions at state x and time k
$\mathscr{A}_{0,k}(x)$	Space of performance-optimal actions at state x and time k, determined by (5.36)
$\mathscr{A}_{m,k}(x)$	Space of mth-bias optimal actions at state x and time k, determined by (5.37)
β	Discount factor
$\mathscr{C}_{m,k}$	Space of mth-bias optimal decision rules at time k, $\mathscr{C}_{m,k} = \prod_{x \in \mathscr{S}_k} \mathscr{A}_{m,k}(x)$, $m = 0, 1, \ldots$.
$\mathscr{C}_m$	Space of mth-bias optimal policies, determined by (5.35), $\mathscr{C}_m = \prod_{k=0}^{\infty} \mathscr{A}_{m,k} = \{u \in \mathscr{C}_{m-1} : A_k^u g_k^{(m+1),u^*} = A_k^{u^*} g_k^{(m+1),u^*}, \forall k\}$,
d	Number of all confluent classes at all times

X.-R. Cao, *Foundations of Average-Cost Nonhomogeneous Controlled Markov Chains*, SpringerBriefs in Control, Automation and Robotics,
https://doi.org/10.1007/978-3-030-56678-4

d_k	Number of confluent classes at time k, $\lim_{k\to\infty} d_k = d$
$\mathscr{D}$	Space of all admissible policies
$\mathscr{D}_n$	Space of all nth-bias optimal policies, $n = 0, 1, \ldots$, with $\mathscr{D}_0$ being the space of all long-run average optimal policies
$\bar{\mathscr{D}}_n$	Space determined by (5.27), $\bar{\mathscr{D}}_n = \{u \in \mathscr{D}_n : A_k^{u^*} g_k^{(n+1),u^*} = \max_{u\in\mathscr{D}_n}[A_k^u g_k^{(n+1),u^*}], \forall k\}$
$E, E^{\mathscr{P}}, E^u$	Expectation, expectation with measure $\mathscr{P}$, and expectation with $\mathscr{P}^u$
$e = (1, \ldots, 1)^T$	A column vector with all components being one
$\eta_{\cdot r}$	Long-run average of the rth confluent class
$\eta_k(x)$	Long-run average starting from state x at time k, $\eta_k(x) = \liminf_{K\to\infty} \frac{1}{K} E\{\sum_{l=k}^{k+K-1} f_l(X_l) \mid X_k = x\}$.
$\mathscr{F}$	σ-field on Ω
$f_k(x), x \in \mathscr{S}_k$	Reward function at k, and the xth component of the reward vector f_k
$f_k := (f_k(1), \ldots, f_k(S_k))^T$	Reward vector at k
$\boldsymbol{f}$	$\boldsymbol{f} := \{f_0, f_1, \ldots, f_k, \ldots\}$
$g_k = (g_k(1), \ldots, g_k(S_k))^T$	Performance potential vector at time k
$g_k(x)$	Performance potential function at time k
$\hat{g}_k(x)$	Bias at state x and time k
$g_k^{(N)}(x)$	Nth bias at state x and time k
$\gamma_k(x, y)$	Relative performance potential at time k
$\gamma_k^{(n)}(x, y)$	Relative nth bias at time k

κ	Hajnal weakly ergodic coefficient
ν	Coefficient of confluencity
Ω	Space generated by $\boldsymbol{X}$, containing all its sample paths
$\mathscr{P}$	Probability measure on $(\Omega, \mathscr{F})$, generated by $\boldsymbol{X}$
P_k	State transition probability matrix of $\boldsymbol{X}$ at time k, which may be non-square, $P_k = [P_k(y\|x)]$, $x \in \mathscr{S}_k$, $y \in \mathscr{S}_{k,out}$
$p_{k,r}(x)$	Probability that a branching state x at time k eventually reaches the rth confluent class
$\mathbb{P}$	Transition law, $\mathbb{P} := \{P_0, P_1, \ldots P_k, \ldots\}$
$\prod$	Cartesian product
$\mathscr{R}_{k,r}$	rth confluent class of states at time k, $r = 1, 2, \ldots, d_k$
S_k	Number of input states in $\mathscr{S}_k$, $S_k = \|\mathscr{S}_k\|$
$\mathscr{S}$	State space of all states at all times
$\mathscr{S}_k$	State space at k, or the input set at time k
$\mathscr{S}_{k,out}$	Output set at k
$\mathscr{T}_k$	Set of branching states at time k
$\tau_k(x, y)$	Confluent time, the time required for two independent sample paths starting from $X_k = x$ and $X'_k = y$ to meet for the first time after k
$\tau_{k,\mathscr{R}}(x)$	The time required for a branching state x at time k to reach a confluent class
u	A policy, $u = (\mathbb{P}, f)$, $u = \{\alpha_0, \alpha_1, \ldots\}$
u^*	An optimal policy
$\upsilon_\beta(x)$	Discount performance measure, $\upsilon_\beta(x) = E\{\sum_{l=0}^{\infty} \beta^l f_l(X_l) \| X_0 = x\}$, $x \in \mathscr{S}_0$, $0 < \beta < 1$

w_k	Bias at time k, $w_k = g_k^{(1)}$
$\boldsymbol{w}_k := (w_k(1), \ldots, w_k(S_k))^T$	Bias vector at time k
X_k	State at discrete time $k = 0, 1, \ldots$
$\boldsymbol{X}$	A Markov chain: $\boldsymbol{X} = \{X_0, X_1, \ldots, X_k, \ldots\}$
$\rightarrowtail\!\!\!\!\leftarrowtail$	$x \rightarrowtail\!\!\!\!\leftarrowtail y$: States x and y are connected

References

Alfa AS, Margolius BH (2008) Two classes of time inhomogeneous Markov chains: analysis of the periodic case. Ann Oper Res 160:121–137

Altman E (1999) Constrained Markov decision processes. Chapman & Hall/CRC, Boca Raton

Anthonisse JM, Tijms H (1977) Exponential convergence of products of stochastic matrices. J Math Anal Apl 59:360–364

Arapostathis A, Borkar VS, Fernandez-Gaucherand E, Ghosh MK, Marcus SI (1993) Discrete-time controlled Markov processes with average cost criterion: a survey. SIAM J Control Optim 31:282–344

Åström KJ (1970) Introduction to stochastic control theory. Academic Press

Bean JC, Smith RL (1990) Denumerate state nonhomogeneous Markov decision processes. J Math Anal Appl 153:64–77

Benaïm M, Bouguet F, Cloez B (2017) Ergodicity of inhomogeneous Markov chains through asymptotic pseudotrajectories. Ann Appl Probab. Institute of mathematical statistics (IMS) 27:3004–3049

Bensoussan A, Frehse J, Yam P (2013) Mean field games and mean field type control theory. Springer, Berlin

Bertsekas DP (2007) Dynamic programming and optimal control, volumes I and II. Massachusetts, Athena Scientific, Belmont, p 2001

Bhawsar Y, Thakur G, Thakur RS (2014) User recommendation system using Markov model in social networks. Int J Comput Appl 86:33–39

Blackwell D (1945) Finite nonhomogeneous Markov chains. Ann Math 46:594–599

Brémaud P (1999) Eigenvalues and nonhomogeneous Markov chains. Markov Chains, vol 31. Gibbs Fields, Monte Carlo Simulation, and Queue, Texts in applied mathematics. Springer, New York, pp 195–251

Briggs A, Sculpher MJ (1998) An introduction to Markov modelling for economic evaluation. PharmacoEconomics 13:397–409

Bryson AE, Ho YC (1969) Applied optimal control: optimization, estimation, and control. Blaisdell, Waltham, Massachusetts

Caines PE (2019) Mean field Games. In: Baillieul J, Samad T (eds) Encyclopedia of Systems and control. Springer, London

X.-R. Cao, *Foundations of Average-Cost Nonhomogeneous Controlled Markov Chains*, SpringerBriefs in Control, Automation and Robotics,
https://doi.org/10.1007/978-3-030-56678-4

Cao XR (2020) On optimization of history-cependent policies of time nonhomogeneous markov chains, Manuscript

Cao XR (2020a) Relative optimization of continuous-time and continuous-state stochastic systems. Springer, Berlin

Cao XR (2020b) Perturbation analysis of steady-state performance and relative optimization. In: Baillieul J, Samad T (eds) Encyclopedia of systems and control. Springer, London

Cao XR (2019) State classification and multi-class optimization of continuous-time and continuous-ctate Markov processes. IEEE Trans Autom Control 64:3632–3646

Cao XR (2019a) The Nth bias and Blackwell optimality of time-nonhomogeneous Markov chains, manuscript

Cao XR (2017) Relative time and stochastic control with non-smooth features. IEEE Trans Autom Control 62:837–852

Cao XR (2016) State classification of time-nonhomogeneous Markov chains and average reward optimization of multi-chains. IEEE Trans Autom Control 61:3001–3015

Cao XR (2015) Optimization of average rewards of time-nonhomogeneous Markov chains. IEEE Trans Autom Control 60:1841–1856

Cao XR (2009) Stochastic learning and optimization - a sensitivity-based approach. (IFAC) Ann Rev Control, invited review 33:11–24

Cao XR (2007) Stochastic learning and optimization - a sensitivity-based approach. Springer, Berlin

Cao XR (2000) A unified approach to Markov decision problems and performance sensitivity analysis. Automatica 36:771–774

Cao XR (1994) Realization probabilities: the dynamics of queueing systems. Springer, New York

Cao XR, Wang DX, Qiu L (2014) Partially observable Markov decision processes and separation principle. IEEE Trans Autom Control 59:921–937

Cao XR, Chen HF (1997) Potentials, perturbation realization, and sensitivity analysis of Markov processes. IEEE Trans Autom Control 42:1382–1393

Cao XR, Zhang JY (2008) Event-based optimization of Markov systems. IEEE Trans Autom Control 53:1076–1082

Cao XR, Zhang JY (2008) The nth-order bias optimality for multi-chain Markov decision processes. IEEE Trans Autom Control 53:496–508

Cassandras CG, Lafortune S (2008) Introduction to discrete event systems, 2nd edn. Springer, Berlin

Chan KS (1989) A note on the geometric ergodicity of a Markov chain. Adv Appl Probab 21:702–704

Chevalier PY, Gusev VV, Hendrickx JM, Jungers RM (2017) Sets of stochastic matrices with converging products: bounds and complexity, Manuscript

Cohn H (1972) Limit theorems for sums of random variables defined on finite inhomogeneous Markov chains. Ann Math Stat 43:1283–1292

Cohn H (1976) Finite non-homogeneous Markov chains: asymptotic behaviour. Adv Appl Probab 8:502–516

Cohn H (1982) On a class of non-homogeneous Markov chains. Math Proc Camb Philos Soc 92:524–527

Cohn H (1989) Products of stochastic matrices and applications. Int J Math Math Sci 12(2):209–233

Connors DP, Kumar PR (1989) Simulated annealing type Markov chains and their order balance equations. SIAM J Control Optim 27:1440–1462

Connors DP, Kumar PR (1988) Balance of recurrence order in time-inhomogeneous Markov Chains with application to simulated annealing. Probab Eng Inf Sci 2:157–184

Dekker R, Hordijk A (1988) Average, sensitive and Blackwell optimal policies in denumerable Markov decision chains with unbounded rewards. Math Op Res 13:395–421

Doeblin W (1937) Exposé de la theorie des chaines simples constantes de Markov à un nombre finid'états. Rev Math de l'Union Interbalkanique 2:77–105

Daubechies I, Lagarias JC (1992) Sets of matrices all infinite products of which converge. Linear Algebra Appl 161:227–263

Douc R, Moulines E, Rosenthal JS (2004) Quantitative bounds on convergence of time-inhomogeneous Markov chains. Ann Appl Probab 14:1643–C1665

Feinberg EA, Shwartz A (eds) (2002) Handbook of Markov decision processes: methods and application. Kluwer Academic Publishers, Boston

Fu MC, Hu JQ (1997) Conditional Monte Carlo: gradient estimation and optimization applications. Kluwer Academic Publishers, Boston

Glasserman P (1991) Gradient estimation via perturbation analysis. Kluwer Academic Publishers, Boston

Griffeath D (1975) A maximal coupling for Markov chains. Z Wahrscheinlichkeitstheorie verw Gebiete 31:95–106

Griffeath D (1975) Uniform coupling of non-homogeneous Markov chains. J Appl Probab 12:753–762

Griffeath D (1976) Partial coupling and loss of memory for Markov chains. Ann Probab 4:850–858

Guo XP, Hernández-Lerma O (2009) Continuous-time markov decision processes. Springer, Berlin

Guo XP, Song XY, Zhang JY (2009) Bias optimality for multichain continuous-time Markov decision processes. Oper Res Lett 37:317–321

Hajnal J (1958) Weak ergodicity in nonhomogeneous Markov chains. Proc Camb Phylos Soc 54:233–246

Hernández-Lerma O, Lasserre JB (1996) Discrete-time Markov control processes - basic optimality criteria. Springer, Berlin

Hernández-Lerma O, Lasserre JB (1999) Further topics on discrete-time Markov control processes. Springer, Berlin

Hinderer K (1970) Foundation of non-stationary dynamic programming with discrete time parameter. Lecture notes in operations research and mathematical systems - economics, computer science, information and control. Springer, Berlin

Ho YC, Cao XR (1991) Perturbation analysis of discrete-event dynamic systems. Kluwer Academic Publisher, Boston

Hopp WJ, Bean JC, Smith RL (1987) A new optimality criterion for nonhomogeneous Markov decision processes. Oper Res 35:875–883

Hordijk A, Yushkevich AA (1999) Blackwell optimality in the class of stationary policies in Markov decision chains with a Borel state and unbounded rewards. Math Methods Oper 49:1–39

Hordijk A, Yushkevich AA (1999) Blackwell optimality in the class of all policies in Markov decision chains with a Borel state and unbounded rewards. Math Methods Oper 50:421–448

Hordijk AA, Lasserre JB (1994) Linear programming formulation of MDPs in countable state space: the multichain case. Oper Res 40:91–108

Jasso-Fuentes H, Hernández-Lerma O (2009a) Ergodic control, bias, and sensitive discount optimality for Markov diffusion processes. Stoch Anal Appl 27:363–385

Jasso-Fuentes H, Hernández-Lerma O (2009b) Blackwell optimality for controlled diffusion processes. J Appl Prob 46:372–391

Kleinrock L (1975) Queueing systems, volume I: theory. Wiley, New York

Kolmogoroff AN (1936) Zur Theorie der Markoffschen Ketten. Math Ann 112:155–160. Selected Works of Kolmogorov AN 2 (Shiryaev AN (ed)). Probability Theory and Math Statistics. Kluwer Acad. Publ

Lewis ME, Puterman ML (2000) A note on bias optimality in controlled queueing systems. J Appl Probab 37:300–305

Lewis ME, Puterman ML (2001) A probabilistic analysis of bias optimality in unichain Markov decision processes. IEEE Trans Autom Control 46:96–100

Lewis ME, Puterman ML (2002) Bias optimality, Handbook of Markov decision processes, 89–111. Kluwer

Li QL (2010) Constructive computation in stochastic models with applications - the RG factorizations. Tsinghua University Press and Springer

Madsen RW, Isaacson DL (1973) Strongly ergodic behavior for non-stationary Markov processes. Ann Probab 1:329–335

Margolius BH (2008) The matrices R and G of matrix analytic methods and the time-inhomogeneous periodic quasi-birth death process. Queueing Syst 60:131–151

Meyn SP, Tweedie RL (2009) Markov chains and stochastic stability, 2nd edn. Cambridge University Press, Cambridge

Mukhamedov F (2013) The Dobrushin ergodicity coefficient and the ergodicity of noncommutative Markov chains. J Math Anal Appl 408:364–373

Park Y, Bean JC, Smith RL (1993) Optimal average value convergence in nonhomogeneous Markov decision processes. J Math Anal Appl 179:525–536

Pitman JW (1976) On coupling of Markov chains. Z Wahrscheinlichkeitstheorie verw Gebiete 35:315–322

Platis A, Limnios N, Le Du M (1998) Hitting time in a finite nonhomogeneous Markov chain with applications. Appl Stoch Models Data Anal 14:241–253

Proskurnikov AV, Tempo R (2017) A tutorial on modeling and analysis of dynamic social networks, Part I. Ann Rev Control 43:65–79

Puterman ML (1994) Markov decision processes: discrete stochastic dynamic programming. Wiley

Roberts GO, Rosenthal JS (2004) General state space Markov chains and MCMC algorithms. Probab Surv 1:20–71

Saloff-Coste L, Zuniga J (2010) Merging and stability for time inhomogeneous finite Markov chains, arXiv:1004.2296

Saloff-Coste L, Zuniga J (2007) Convergence of some time inhomogeneous Markov chains via spectral techniques. Stoch Process Appli 117:961–979

Sericola B (2013) Markov chains: theory, algorithms and applications. Wiley

Sonin IM (1991) An arbitrary nonhomogeneous Markov chain with bounded number of states may be decomposed into asymptotically noncommunicating components having the mixing property. Theory Probab Appl 36:74–85

Sonin IM (1996) The Asymptotic Behaviour of a General Finite Nonhomogeneous Markov Chain (the Decomposition-Separation Theorem), Statistics, Probability and Game Theory, vol 30. Lecture Notes - Monograph Series

Sonin IM (2008) The decomposition-separation theorem for finite nonhomogeneous Markov chains and related problems. In: Ethier SN, Feng J, Stockbridge RH (eds) Markov processes and related topics: A Festschrift for Thomas G. Kurtz

Teodorescu D (1980) Nonhomogeneous Markov chains with desired properties - the nearest Markovian model. Stoch Process Appl 10:255–270

Veinott AF (1969) Discrete dynamic programming with sensitive discount optimality criteria. Ann Math Stat 40:1635–1660

Xia L, Jia QS, Cao XR (2014) A tutorial on event-based optimization - a new optimization framework. Discrete Event Dyn Syst Theory Appl, Invited paper 24:103–132

Xia L, Guo XP, Cao XR (2020) On the existence of optimal stationary policies for average Markov decision processes with countable states, manuscript submitted

Zhang JY, Cao XR (2009) Continuous-time Markov decision processes with Nth-bias optimality criteria. Automatica 45:1628–1638

Zheng Z, Honnappa H, Glynn PW (2018) Approximating performance measures for slowly changing non-stationary Markov chains, arXiv: 1805.01662v1

Zeifman AI, Isaacson DL (1994) On strong ergodicity for nonhomogeneous continuous-time Markov chains. Stoch Process Appl 50:263–273

Series Editors' Biographies

Tamer Başar is with the University of Illinois at Urbana-Champaign, where he holds the academic positions of Swanlund Endowed Chair, Center for Advanced Study (CAS) Professor of Electrical and Computer Engineering, Professor at the Coordinated Science Laboratory, Professor at the Information Trust Institute, and Affiliate Professor of Mechanical Science and Engineering. He is also the Director of the Center for Advanced Study—a position he has been holding since 2014. At Illinois, he has also served as Interim Dean of Engineering (2018) and Interim Director of the Beckman Institute for Advanced Science and Technology (2008–2010). He received the B.S.E.E. degree from Robert College, Istanbul, and the M.S., M.Phil., and Ph.D. degrees from Yale University. He has published extensively in systems, control, communications, networks, optimization, learning, and dynamic games, including books on non-cooperative dynamic game theory, robust control, network security, wireless and communication networks, and stochastic networks, and has current research interests that address fundamental issues in these areas along with applications in multi-agent systems, energy systems, social networks, cyber-physical systems, and pricing in networks.

In addition to his editorial involvement with these Briefs, Başar is also the Editor of two Birkhäuser series on *Systems & Control: Foundations & Applications* and *Static & Dynamic Game Theory: Foundations & Applications*, the Managing Editor of the *Annals of the International Society of Dynamic Games* (ISDG), and member of editorial and advisory boards of several international journals in control, wireless networks, and applied mathematics. Notably, he was also the Editor-in-Chief of *Automatica* between 2004 and 2014. He has received several awards and recognitions over the years, among which are the Medal of Science of Turkey (1993); Bode Lecture Prize (2004) of IEEE CSS; Quazza Medal (2005) of IFAC; Bellman Control Heritage Award (2006) of AACC; Isaacs Award (2010) of ISDG; Control Systems Technical Field Award of IEEE (2014); and a number of international honorary doctorates and professorships. He is a member of the US National Academy of Engineering, a Life Fellow of IEEE, Fellow of IFAC, and Fellow of SIAM. He

X.-R. Cao, *Foundations of Average-Cost Nonhomogeneous Controlled Markov Chains*, SpringerBriefs in Control, Automation and Robotics,
https://doi.org/10.1007/978-3-030-56678-4

has served as an IFAC Advisor (2017–), a Council Member of IFAC (2011–2014), president of AACC (2010–2011), president of CSS (2000), and founding president of ISDG (1990–1994).

Miroslav Krstic is Distinguished Professor of Mechanical and Aerospace Engineering, holds the Alspach endowed chair, and is the founding director of the Cymer Center for Control Systems and Dynamics at UC San Diego. He also serves as Senior Associate Vice Chancellor for Research at UCSD. As a graduate student, Krstic won the UC Santa Barbara best dissertation award and student best paper awards at CDC and ACC. Krstic has been elected Fellow of IEEE, IFAC, ASME, SIAM, AAAS, IET (UK), AIAA (Assoc. Fellow), and as a foreign member of the Serbian Academy of Sciences and Arts and of the Academy of Engineering of Serbia. He has received the SIAM Reid Prize, ASME Oldenburger Medal, Nyquist Lecture Prize, Paynter Outstanding Investigator Award, Ragazzini Education Award, IFAC Nonlinear Control Systems Award, Chestnut textbook prize, Control Systems Society Distinguished Member Award, the PECASE, NSF Career, and ONR Young Investigator awards, the Schuck ('96 and '19) and Axelby paper prizes, and the first UCSD Research Award given to an engineer. Krstic has also been awarded the Springer Visiting Professorship at UC Berkeley, the Distinguished Visiting Fellowship of the Royal Academy of Engineering, and the Invitation Fellowship of the Japan Society for the Promotion of Science. He serves as Editor-in-Chief of *Systems & Control Letters* and has been serving as Senior Editor for *Automatica* and *IEEE Transactions on Automatic Control*, as editor of two Springer book series—*Communications and Control Engineering* and *SpringerBriefs in Control, Automation and Robotics*—and has served as Vice President for Technical Activities of the IEEE Control Systems Society and as chair of the IEEE CSS Fellow Committee. Krstic has coauthored thirteen books on adaptive, nonlinear, and stochastic control, extremum seeking, control of PDE systems including turbulent flows, and control of delay systems.

Index

A
Absorbing state, 20
Action, 6
Asynchronicity, 75
Asynchronous sequences, 73

B
Bias, 6, 44
Bias difference formula, 50
Bias optimality conditions, 50
    necessary, 51
Bias potential, 48
Blackwell optimality, 102
Branching state, 20

C
Coefficient, 16
    Hajnal weak ergodicity, 16
    of confluencity, 16
Confluencity, 1, 14
    strong, 14
Confluent state, 20
Confluent time, 14
Connectivity, 19
    strong, 19

D
Discounted reward, 100
Dynkin's formula, 36

G
Geometrically ergodic, 81, 107

I
Infinitesimal generator, 36, 62
Inf-subsequence, 30

L
Laurent series, 100
Lim-subsequence, 30
Linear Quadratic Gaussian (LQG) control, 3
Long-run average, 29

M
Markov chains
    multi-class, 26
    strongly connected, 24, 61
    uni-chain, 26

N
*N*th bias, 83
*N*th-bias difference formula, 86

O
Optimality conditions, 38, 39
    for bias, 50
    for long-run average, 38, 39
    for *N*th bias, 98
    multi-class, necessary, 70
    multi-class, sufficient, 69

P
Performance difference formula

X.-R. Cao, *Foundations of Average-Cost Nonhomogeneous Controlled Markov Chains*, SpringerBriefs in Control, Automation and Robotics,
https://doi.org/10.1007/978-3-030-56678-4

for average reward, 38
for bias, 50
for *N*th bias, 86
Performance measure, 6
Performance potential, 34
multi-class, 62
Periodic Forecast Horizon (PFH) optimality, 11, 80
Poisson equation
for bias, 45, 82
for bias potential, 48
for long-run average, 34, 36
for *N*th bias, 83
multi-class, 62, 63
Policy, 7

R
Reachable, 20
Relative bias potential, 48
Relative optimization, 1, 8, 11
Relative performance potential, 32
multi-class, 61
Reward function, 5

S
Sensitivity-based approach, v
Sensitivity (or N) discount optimality, 104
State space, 4
input set, 4
output set, 4

T
Time-nonhomogeneous Markov chains, 4
Transition law, 4
Transition probability matrix, 4

U
Under-selectivity, 8, 39
Uniformly positive, 40

W
Weak ergodicity, 10, 15
Weakly recurrent, 41

Printed in the United States
By Bookmasters